THE OFFICIAL BOOK OF CIRCULAR SUDOKU

THE OFFICIAL BOOK OF
CIRCULAR SUDOKU

BOOK I

120 PUZZLES

Peter M. Higgins and Caroline Higgins

A PLUME BOOK

PLUME
Published by Penguin Group
Penguin Group (USA) Inc., 375 Hudson Street, New York, New York 10014, U.S.A.
Penguin Group (Canada), 90 Eglinton Avenue East, Suite 700, Toronto, Ontario,
Canada M4P 2Y3 (a division of Pearson Penguin Canada Inc.)
Penguin Books Ltd., 80 Strand, London WC2R 0RL, England
Penguin Ireland, 25 St. Stephen's Green, Dublin 2, Ireland (a division of Penguin Books Ltd.)
Penguin Group (Australia), 250 Camberwell Road, Camberwell, Victoria 3124, Australia
(a division of Pearson Australia Group Pty. Ltd.)
Penguin Books India Pvt. Ltd., 11 Community Centre, Panchsheel Park,
New Delhi – 110 017, India
Penguin Books (NZ), cnr Airborne and Rosedale Roads, Albany, Auckland 1310,
New Zealand (a division of Pearson New Zealand Ltd.)
Penguin Books (South Africa) (Pty.) Ltd., 24 Sturdee Avenue, Rosebank,
Johannesburg 2196, South Africa

Penguin Books Ltd., Registered Offices: 80 Strand, London WC2R 0RL, England

First published by Plume, a member of Pengin Group (USA) Inc.

First printing, July 2006
1   3   5   7   9   10   8   6   4   2

Copyright © Peter Higgins, 2006
All rights reserved

Ⓟ REGISTERED TRADEMARK—MARCA REGISTRADA

ISBN 0-452-28796-0

Printed in the United States of America
Set in Adobe Caslon • Designed by Elke Sigal

Contents

Introduction

Sudoku: The Next Generation

Nothing has united the peoples of the planet over the last year like the Sudoku craze. First appearing in the United States in *Dell* magazine, Sudoku became a worldwide phenomenon only after undergoing a Japanese makeover and emerging in its present form under the name *Su Doku* (literally, single number).

Although the popularity of the nine-by-nine grid continues unabated, there have been attempts at variations. Most of these involve larger and more complicated versions of the same grid, or they depart from the pure nature of Sudoku through hybrid inventions that introduce a mixture of arithmetic and Sudoku placement rules. Circular Sudoku, on the other hand, has as its basis a pure Sudoku game—the set of symbols used does not matter, for they can just as easily be letters, or signs of the zodiac, or even colors. In this way Circular Sudoku continues the Sudoku tradition in that the puzzle is one of pure logic, meaning that its unique solution is one displaying a certain perfect balance of symbols. The overall pattern is all that matters, and this pattern is revealed, step-by-step, through a process of elimination by posing exactly the right question at each stage.

The first Circular Sudoku puzzle appeared in the British newspaper *The Sunday Telegraph* on June 26, 2005. Since then it has featured in a number of other papers, magazines, and books. This is the first book dedicated to Circular Sudoku.

The basic Circular Sudoku puzzle has rules that can be expressed in one sentence:

Every symbol appears in each ring and each pair of adjacent "slices of the pie."

In a four-ring puzzle this means that each of the numbers 1–8 must appear once in each ring and in every pair of touching slices. Armed with this one instruction, the reader is invited to dive in and tackle the first puzzle. It really is not very hard—the first puzzles in the book should take less time than a regular Sudoku grid.

Below you will find the basic techniques explained, but you might prefer to ignore all that and head straight for the puzzles themselves. As you flick through the 120 puzzles on offer you will quickly spot that they do not all share the same format. This highlights an advantage of the circular form of the game. It is possible to increase the size of the array from four rings to five or even six rings without the puzzle becoming too complicated.

Basic four-ring puzzles are the simplest, and they make up the first section of the book. Next we graduate to the five-ring version that has 50 cells (as opposed to 32 for the four-ring puzzle) and requires more subtle techniques. Five-ringers make up the second section of the book. The third section is a potpourri of puzzles, some with more than five rings, where additional constraints are added that provide novel twists of their own. There is a different version we call Target Sudoku that involves a four-ring, 48-cell grid. Although not necessarily diabolically difficult,

Target Sudoku does not succumb to the basic techniques that ultimately will sort out any Circular Sudoku puzzle.

We hope this little book, for a few minutes every now and then, brightens up your working life by allowing you the luxury of freeing your mind to take a little refreshing exercise. Have fun!

How to Solve Circular Sudoku Puzzles

Some of you may have successfully tackled some of the puzzles already and so, on the way, will have invented your own techniques for doing so. On reading the methods described, you might then find yourself thinking, "That's not the way that I do it." As long as you are consistently cracking the puzzles, you must be on the right track. All the same, you might pick up a few weapons to add to your arsenal.

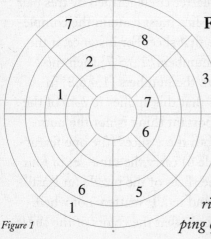

Figure 1

Four-Ring Puzzles (Puzzles 1–35)

Even hardened Sudoku enthusiasts often take fright when they see their first Circular Sudoku puzzle. The rule for a four-ring puzzle (puzzles 1–35) can be expressed as follows:

Each of the numbers 1–8 must appear in each of the four rings and each of the eight overlapping quarter circles. See Figure 1.

Readers are often uneasy about what it is they are supposed to do. The part about each symbol appearing in each ring seems straightforward enough. But the second part of the rule can leave people in a quandary: Each of the eight overlapping quarter circles must contain each number as well. At first sight this appears to make things horribly complicated. However, it is this part of the rule that makes tackling the puzzle possible!

Step 1: The Parallel Breakup

Focus on any slice in the given puzzle—consider the slice that features the number 3. In the solution, this slice will contain a set of four numbers in some order. We do not know immediately what these four numbers are, except that the set does include 3 and 7. Now consider one of the slices next to this one. Since all eight numbers must occur in the quarter circle formed by this pair of slices, it follows that this second slice must contain the complementary set of four numbers to the first slice. For instance, if the first slice ended up with the set of numbers {2,3,5,7} then the slice next to it would have to feature the numbers {1,4,6,8} in some order. The key observation is that this applies to *both* of the slices next to the first one. The important thing is not the numbers themselves but the fact that both these slices (the one featuring the 8 and the other with a single 6) carry the same set of four numbers. Suppose that we color the slices of the puzzle (at least in our imaginations) alternately blue and red. It matters not where you start nor which color you use, only that the colors of the slices alternate as we travel around the puzzle. All slices of the one color feature exactly the same set of numbers (but in different orders).

This provides us with the first step in solving a Circular Sudoku puzzle. Begin with any slice (call it a blue slice) and find the set of all "blue" numbers by listing the numbers that occur in that slice and any of the other slices of the same color, which you find by taking every second slice in the puzzle. Applying this to *Figure 1* above and beginning with the slice featuring the number 3 as our first blue slice we adopt this procedure by going around the circle, let us say, clockwise. The first blue numbers are 3 and 7. The next blue slice contains the number 5, although the third blue slice has no clues for it is empty. However, the final blue slice has the two numbers 7 and 2, giving us as our blue set B = {2,3,5,7}. The remaining "red" set for the other four slices consists of the remaining numbers so that we can write it as R = {1,4,6,8}.

We could have gone after the red symbols first. In this case, only three of them are visible in the puzzle—{1,6,8} as the symbol 4 does not appear as a clue. This tells us straightaway that the missing red number must be 4. Alternatively, we can find the blue numbers first, as we did here, and then the reds are the ones left over.

In any puzzle, at most one number can be totally absent from the given set of clues. For that reason, the blue and red sets can always be found.

Step 2: Solving Each Partial Puzzle

We can therefore complete the puzzle by solving the blue and red halves separately.

Let us write the four cells of each of the blue slices as four rows of four spaces. We begin with the empty slice giving us a blank row. The second blue row features the numbers 7 and 2, and we shall write down a row from left to right as it appears from the outside of the puzzle moving in toward the center, giving us a second row: 7 _ 2 _. Doing likewise with the final two blue slices leaves us with a 4 x 4 square that is partially completed.

By this stage we know what symbols should appear in this square—the blue numbers. Each row of the square corresponds to a slice that must contain all of the blue numbers, so every row must contain every blue number. However, the same applies to each column of the square. Reading down any column of the 4 x 4 grid we are traveling within a single ring of the puzzle (although we are hopping two steps at a time, as we are staying in the blue slices). Since a ring cannot contain any repeats, the same applies to each column of the square. *See Figure 2.*

_	_	_	_		5	7	3	2
7	(3)	2	_		7	3	2	5
3	_	(5)	7		3	2	5	7
(2)	5	_	_		2	5	7	3

Figure 2

A square like this one is known as a Latin square. Part of the charm of Circular Sudoku is the way in which Latin squares have been smuggled into the puzzle. At first glance, squares are nowhere to be seen, but they are there, lurking below the surface.

We have reduced the problem of Circular Sudoku to that of completing a pair of partially completed and interleaved Latin squares—a blue square and a red square. Completing this task completes the puzzle.

To complete the squares, you need to ask yourself:

Question 1: *What numbers can legally occupy this cell?*
The key step is to find an empty position in your square for which there is only one answer to the question. In this case the circled entry in position (2,2) (meaning the second row and second column) can only be filled with a 3, for the symbols 7 and 2 already appear in its row and 5 is at the foot of its column. Equally, we could start filling in the square at position
(4,1) as only a 2 can be placed here
or position (3,3), which must
house a 5. As you fill places,
more of the remaining empty
slots will have but one possible
entry and in this way the
entire square is filled, at least
for a puzzle with fewer than
five rings. The unique completion allows us to fill the blue
portion of the puzzle.
See Figure 3.

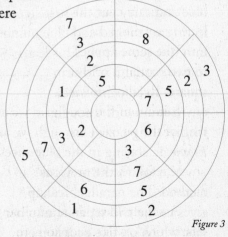

Figure 3

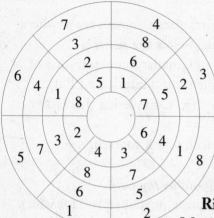

Figure 4

It only remains to solve the red portion of the puzzle in the same fashion. It does not matter which red segment you take as your top row. This leads us to the completed solution that can be seen in *Figure 4*.

Puzzles with Five or More Rings (Puzzles 36–70)

Moving on to five-ring puzzles, the preceding method still applies. A five-ring puzzle has ten segments, making 50 cells in all, and requires ten different symbols, which we may naturally take to be the digits 0–9. Once again we may regard the segments as alternately colored red and blue and carry out the first step of collecting up the red and blue numbers into two sets of five. This leads to two partial 5 x 5 Latin squares to be completed.

Let us work through the five-ring puzzle in *Figure 5*.

We follow the previous method. The parallel breakup gives us the two separate number sets, which on this occasion are

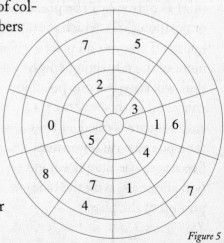

Figure 5

B = {0,2,3,4,7} and R = {1,5,6,8,9}, with the number 9 located by elimination. Setting up the red and blue

							0		
	5					7		2	
(9)	6	1							3
		1			7			4	
	8			5		4	7		

Figure 6

partial Latin squares yields the diagram in *Figure 6*.

By posing Question 1 to the red square, we get an immediate foothold at position (3,2) as that entry must be a 9. From there more of the cells in this square have only one possible number and the square may be completed.

However, turning our one and only weapon on the blue square we seem to be stuck. There are seventeen blank cells and the answer to our Question 1 for each of them gives at least two possibilities in every case. This calls for a new tack, one that you may well have tried for yourself already. Instead of focusing on an empty cell and asking ourselves, What can go here?, we can take one of the symbols and ask . . .

Question 2: *Where can this symbol go?*
A row of the square corresponds to a slice of the puzzle, while as we look down a column we are traveling around the same ring (two steps at a time, as we are working with just one color number).

When applying Question 2 you should look for symbols that have occurred frequently as clues in the puzzle. In this instance we are already given the position of three of the five 7's, and so we may ask where the 7 in the other two rows could go. Asking this question for the first row it would seem that a 7 could possibly go in either of the positions (1,4) or (1,5).

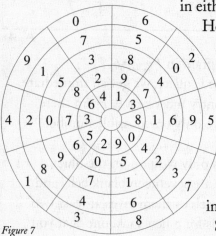

However, there is no such ambiguity for row 3: the only place the 7 can go without a clash is at (3,4), and so we can write it in. Having found the position of all the instances of the symbol 7, the partial Latin square and the puzzle can be completed yielding the solution as pictured in *Figure 7*.

Figure 7

Steps 1 and 2 combined with Questions 1 and 2 are enough to solve any of the basic puzzles with up to five rings. With a little practice, you may well be able to apply this method without drawing up the Latin squares explicitly.

Advanced Strategies (Puzzles 71–100)

The puzzle setter can always depart from or augment the basic rules of the game in any way that makes for a good puzzle. Some puzzles in this book have too few clues to solve just using the basic rule and so another clue is thrown in to help you out. Since

there is potentially no end to these
variants there is not much that
can be said about them in gen-
eral. However, a particular rule
can lead to a surprisingly good
variant of the standard puzzle.

Figure 8

Ring Sudoku

Ring Sudoku is a standard puzzle
but with a twist in that we are told
that every ring is identical, just each is
offset from the other. Since the standard rules apply, we can
apply the standard method and the extra condition will surely
make it easier. Let us see what happens when we try the puzzle
in *Figure 8*.

As usual, the set 0–9 separates into two sets that we call
B = {2,3,4,5,7} and R = {0,1,6,8,9}. We can then set up the blue
and red partial squares, but they look pretty sparse! *See Figure 9.*

Neither Questions 1 nor 2 allow us to start filling empty cells
immediately, for there are too few clues to proceed that way. (No
number appears twice, and 6 is missing entirely.) We have to
make use of the extra hint, which allows for the two squares that
are normally independent of each other, to interact.

Look to position (4,1) of the left-hand square in the diagram.
Clearly the entry must be either 2 or 3. However, it cannot be a 3,
as 3 must be followed by 7—as you read down a column of either
of these squares you are traveling clockwise around a ring two

```
_ _ _ 2 _        _ _ _ _ _
_ _ _ _ 1        _ _ _ _
_ 3 _ _          _ _ _ _ 8
(2) 7 4 _ _      _ _ 9 _ _
5 (2) _ _ _      _ _ _ 0 _
```

Figure 9

steps at a time. Since every ring has the same order, the same holds for the columns of each square.

Hence we can fill in a 2 at the cell in position (4,1). Next, focus on cell (5,2) of the same square, whose entry is either 2 or 4. It again must be a 2 in this position for 2 cannot sit in position (1,2), nor can it be in position (2,2) because we now can see that 2 is followed by 5. Hence 2 lies at the bottom of column 2.

The hard work is now done as we can see that the order of the blue set is 3,7,2,5, and then 4 by simple elimination. We can immediately write down each of the first four columns of the blue square, whereupon the final column can also be completed. We have arrived at the situation depicted in *Figure 11*.

The signed numbers below each column of the blue square show

```
4 5 7 2 3        _ _ _ _ _
3 4 2 5 7     1  _ _ _ _
7 3 5 4 2        _ _ _ _ 8
2 7 4 3 5        _ _ 9 _ _
5 2 3 7 4        _ _ _ 0 _
  +1 +3 +2 +4      +1 +3 +2 +4
```

Figure 10

how many places we move the first column down in order to get the corresponding column. For example, the +2 at the foot of the fourth column indicates that we obtain the fourth column by beginning two places below the top, at position (3,4) and then writing the first column in order: 4,3,7,2,5.

These numbers tell you how to rotate the outside ring to get the inner ones: one place clockwise for the second ring, 3 for the third, and so on. They apply equally well to the red square. This allows us to write down the first column of the red

4	5	7	2	3		9	6	0	8	1
3	4	2	5	7		1	9	8	6	0
7	3	5	4	2		0	1	6	9	8
2	7	4	3	5		8	0	9	1	6
5	2	3	7	4		6	8	1	0	9
	+1	+3	+2	+4			+1	+3	+2	+4

Figure 11

square: The 9 in column 1 is three positions up from the 9 shown at (3,4), so there is a 9 at (1,1); similarly, the 0 in the first column is at position (1,3), two rows up from where we see the 0 in the red square, and to find the position of the 8 we have to wind back four rows starting from row number 3, giving, in order $3 \rightarrow 2 \rightarrow 1 \rightarrow 5 \rightarrow 4$, so that there is an 8 in position (1,4). The missing 6 must then lie at position (1,5).

The first column of the red square, reading from top to bottom, is then 9,1,0,8,6. Repeating this order allows us to write down columns 3, 4, and 5 and finally column 2 is found by elimi-

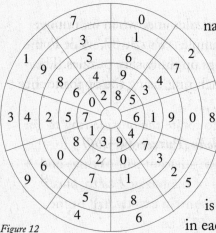

Figure 12

nation. The completed blue and red squares are then as we see in *Figure 11*, which gives the solution as shown in *Figure 12*.

Varying the Ring and Slice Rules

The standard rule for Circular Sudoku that applies to the slices is that all symbols have to appear in each touching pair, and this leads, as we have now seen, to a problem involving two Latin squares. A natural variant of this rule is to insist instead that this constraint applies to each pair of diagonally opposite slices.

This is an appealing rule but is considerably weaker than the standard one. For any Circular Sudoku puzzle with an odd number of rings, the usual rule has as a consequence that the opposite slice rule is also obeyed—each pair of opposite slices will have complementary colors. However, the opposite slice rule does not have as an automatic consequence that all symbols need appear in adjacent slices. A consequence of this is that many more clues must be given than in standard Circular Sudoku. In fact, at least half of the puzzle needs to be provided to ensure a unique solution.

For this reason, opposite slice rule puzzles generally need to be augmented with some other clue to cut down on the number of given cells. For example, good puzzles arise when we impose this rule on Ring Sudoku—an example is Puzzle 87, while Puzzles 74 and 75 are regular four rings opposite slice puzzles. The natural approach is to complete each ring, using the opposite slice rule to narrow down your options.

Symmetric or Double Circular Sudoku (Puzzles 96–100)
In Symmetric Circular Sudoku each adjacent pair of rings and slices must contain all the symbols of the puzzle exactly once.

Each slice and each ring must contain exactly half of the symbols used in the puzzle. *Figure 13* is an example using 6 rings and 6 slices (the minimum necessary for this type of puzzle to work) on a set of 12 symbols that we take to be {A,B,0,1,2, . . . ,9}.

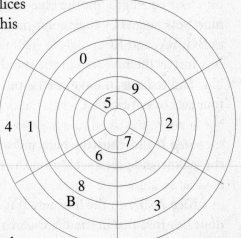

Figure 13

As before, we have the red and blue slices—the alternating red slices will all contain the same set of six symbols while all the blue slices will feature the complementary set of six. But the rings are now subject to exactly the same rule. Each ring contains the opposite set of six to the rings adjacent to it so that the rings also can be thought to have alternating colors, yellow and green, say, corresponding to two sets of six symbols.

It is important to note that the ring colors are not the same as those of the slices! For example, let us examine how a golden ring meets the slices of the puzzle. As we trace around the ring we alternately hit blue and red numbers. It follows that the yellow set will consist of three red and three blue symbols, while the green rings will each feature the other three red and blue numbers.

Just like the standard puzzle breaks up the collection of symbols into two equal sets, in this puzzle type the collection breaks into four equal groups: the blue and yellow set (BY) and three more sets tagged by their color pairs that we may naturally write as BG, RY, and RG.

Just as the first step in the standard puzzle is to identify the blue and red sets, the symmetric puzzle requires us to find these four sets of symbols.

Step 1   Identify the blue and red sets (from the slices) and the yellow and green sets (from the rings).

Step 2   Form the four sets BY, BG, RY, and RG. This is done, for instance in the BY case, by taking those numbers common to both the blue and yellow sets.

In our example above where there are twelve symbols, Step 1 gives us:

B = {0,2,5,6,8,B} R = {1,3,4,7,9,A}
Y = {1,2,3,5,7,B} G = {0,4,6,8,9,A}
Carrying out the second step we find that:
BY = {2,5,B}, BG = {0,6,8}, RY = {1,3,7}, RG = {4,9,A}.

Let us focus on one of them, BY say, as the story is the same for them all. In the case of a six-ring puzzle, there are 3 x 3 = 9 blue-yellow cells: three in each blue slice, and equally, three in each yellow ring. Each blue-yellow symbol has to appear once in each blue slice and once in each yellow ring. In other words, these nine cells form a little 3 x 3 Latin square. Partial 3 x 3 Latin squares are easy to complete—once you know any two symbols, not in the same row or column, finding the unique solution is child's play. The final stage is:

Step 3   Identify each of the four minor Latin squares.
We shall do this in detail for the red-green square, which is perhaps the trickiest as only two of the three symbols in RG appear in the puzzle.
Figure 14 shows the red-green Latin square and its completion.

4	__	__		4	9	A
__	__	9		A	4	9
__	__	__		9	A	4

Figure 14

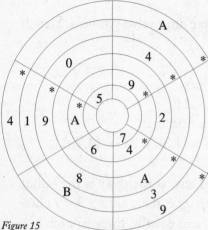

The partially solved puzzle in which the red-green part is completed is shown in *Figure 15*, where we have marked each red-green cell with an asterisk. You now have only to repeat this procedure for the other three mini-squares to finish the job.

Figure 15

Target Sudoku (Puzzles 101–120)

The following is a more difficult version of Circular Sudoku known as Target Sudoku, as the alternating pattern of colored slices is reminiscent of a dartboard or an archery target. In the 48-cell puzzle there are twelve different symbols; we have used A,B,0,1,2, . . . ,9. The rule of the game is:

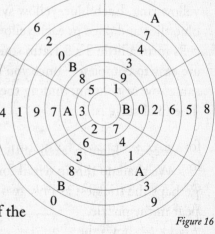

Figure 16

Each symbol must appear in each of the four rings and each of the six white–black–white quarter circles.

Our usual blue-red separation method that was relevant to our puzzles up to this point now goes out the window, as the puzzle simply does not split into two. There are special properties that emerge nonetheless. A consequence of the rules is that:

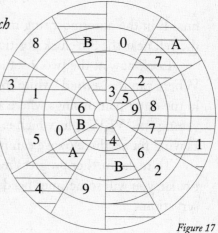

Figure 17

1. Every symbol must appear twice in a white cell and twice in a black cell.
2. Any symbol that appears three times in the puzzle has its final place fixed.

This second rule gives us a way to get started in this example, for if the position of symbol *x* is given three times then its final fourth entry is easy to spot. Just focus on the remaining ring that does not yet have an *x*. The positions of the three other *x*'s allow you to determine where that final *x* must lie in this ring.

For example, in the puzzle above, B appears three times so the remaining place may be determined by the WBW rule. The final B in the outermost ring must occupy a white cell and only the cell near the 3 o'clock position obeys the rules.

Look now at the two instances of 6 given in the puzzle. Where can the second "white" 6 lie? There is only one white cell in the outer pair of rings where a 6 can be placed without contravening the WBW rule, which is at 1 o'clock in the outer ring. Having found three 6's we can now place the fourth 6 as well.

To make further progress requires a little more guile—in particular, you will sometimes need to make use of the following observation:

3. The four instances of any symbol appear in either
   (a) a pair of diametrically opposite white slices and black slices at right angles to one another, or
   (b) a pair of white slices separated by a black-white-black quadrant and a pair of black slices separated by a single white slice.

In our example, focus on the two instances of 1, which are both "black." According to 3(a) the two "white" instances of 1 occur in the white slices at 1 o'clock and 7 o'clock in the inner two rings. The only way that can be accommodated is shown in *Figure 18*. Moreover, this leaves a WBW quadrant in need of a 4 that has only one empty cell. Filling in that 4 gives us three of the 4's whereupon the placement of the final 4 is determined.

The two given 0's on the other hand are white and their slices are separated by 120 degrees. In accord with 3(b), the two black

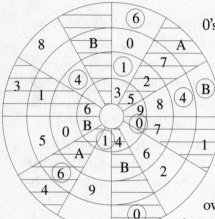

Figure 18

0's will lie in the black slices at 3 and 5 o'clock, respectively, and must occupy the inner and outermost rings. There is only one way of doing this, so we can complete the positioning of the 0's.

In this fashion, you will sometimes need to consider the overall placement of all four instances of a particular symbol in order to make progress. *Figure 18* circles the nine deductions we have made so far.

Armed with this bag of tricks, you should be able to overcome all the Target Sudoku puzzles in this book. The complete solution to this one is pictured below.

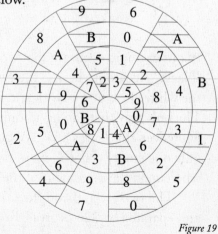

Figure 19

Easy Does It

Four-Ring Puzzles

The rule:

The numbers 1–8 are to appear in each of the four rings
and each pair of adjacent slices of the circle.

Puzzle 1 It all started with four-ring puzzles like this one: the numbers 1–8 are to appear in each of the four rings and each of the eight quarter circles.

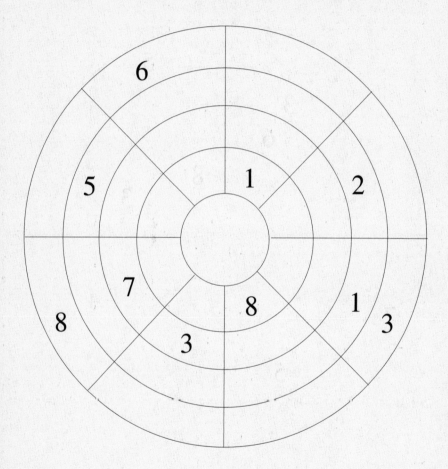

Puzzle 2   This four-ring puzzle has only nine given clues, yet the solution is still unique.

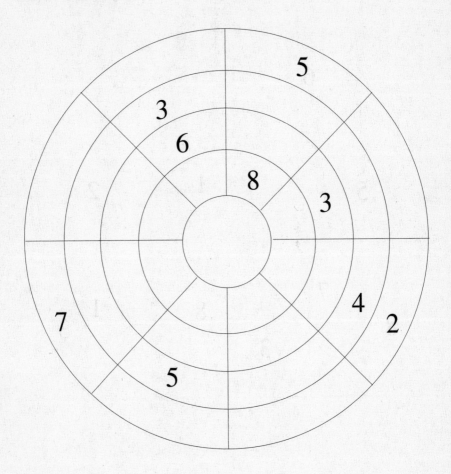

Puzzle 3   A simple donut to get you started.

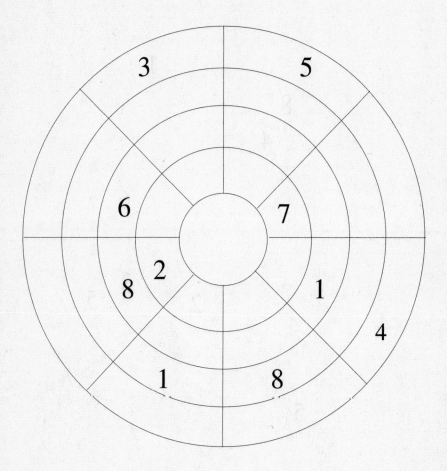

Puzzle 4   Just a jumble.

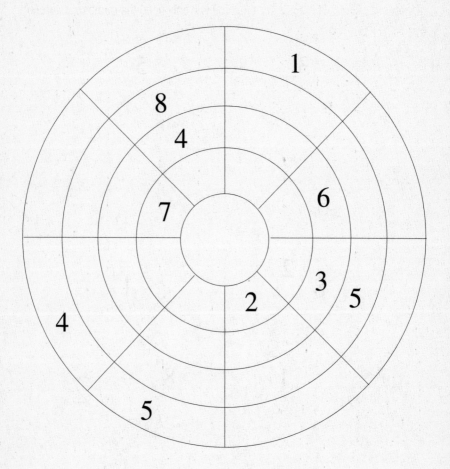

Puzzle 5 A three-ring puzzle—not too difficult as you would expect! The rules are unchanged: each symbol is to appear in each ring and each pair of neighboring slices. Of course this time you will only need the numbers 1 through to 6 to fill it out.

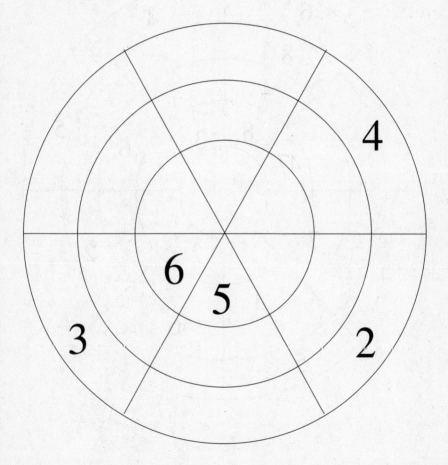

Puzzle 6 This time the clues spiral in.

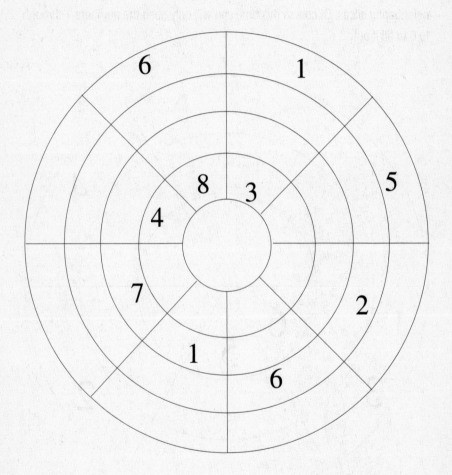

Puzzle 7   See if you can tackle one with an empty segment.

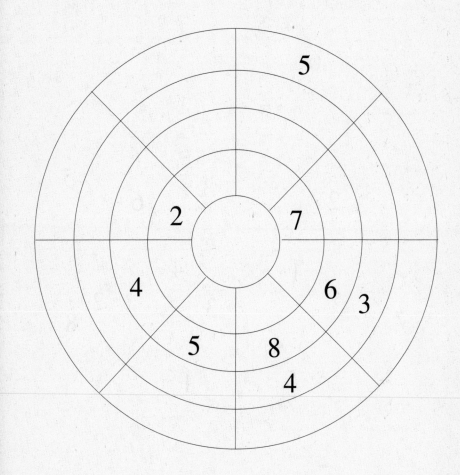

Puzzle 8 This time one large slice of the pie is missing—one entire quarter circle is empty!

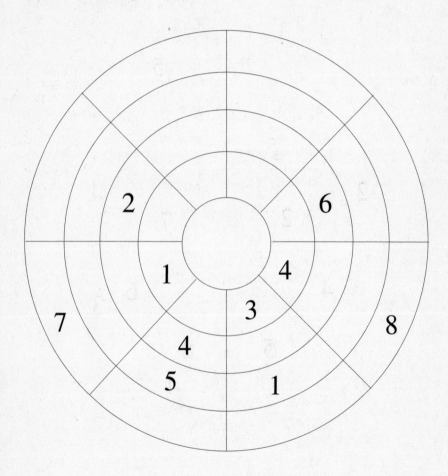

Puzzle 9   Try to solve this one, which has a vacant outer ring.

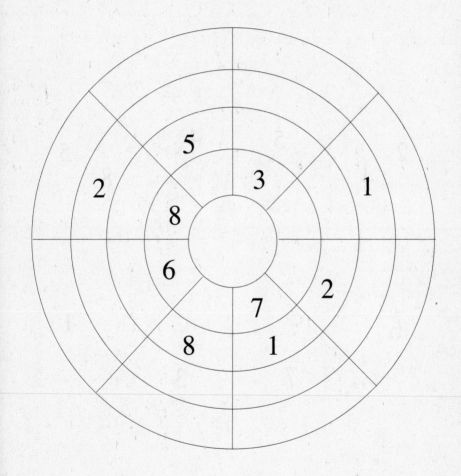

Puzzle 10 Two horizontal lines—nothing could be simpler?

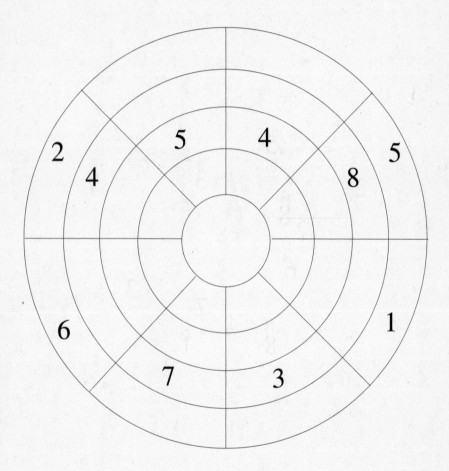

Puzzle 11 No pattern, just a big jumbly mess.

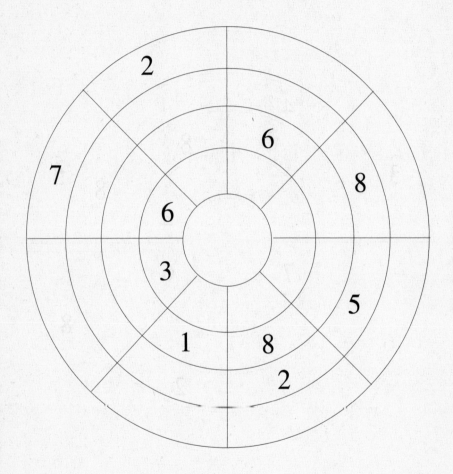

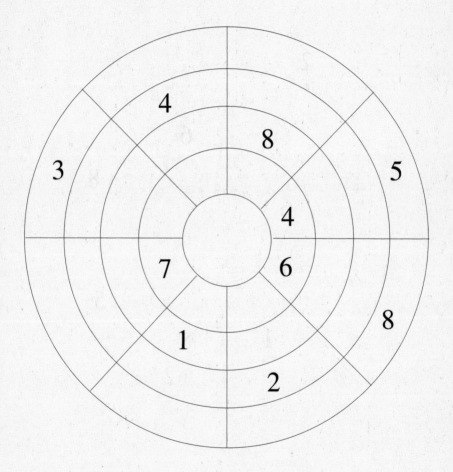

Puzzle 13 This was the first ever Circular Sudoku puzzle.

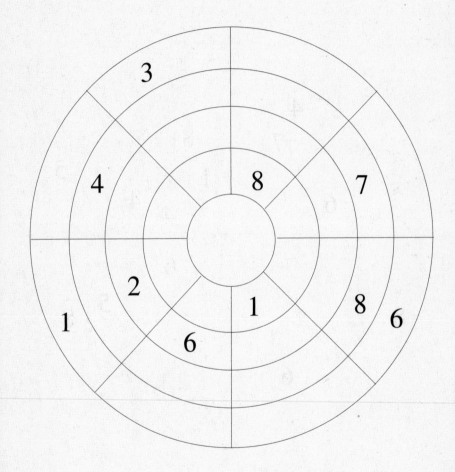

Puzzle 14 Again, just the nine numbers—you can't have fewer clues than this.

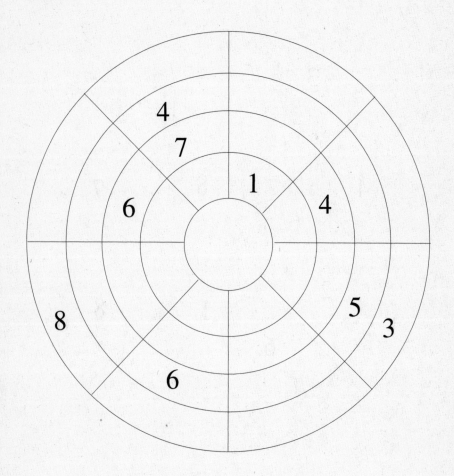

Puzzle 15 Are they getting easier?

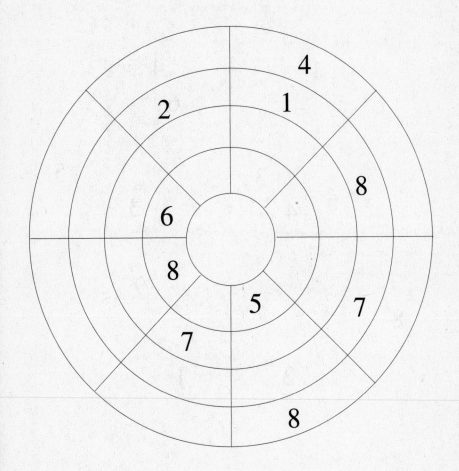

Puzzle 16 More standard fare.

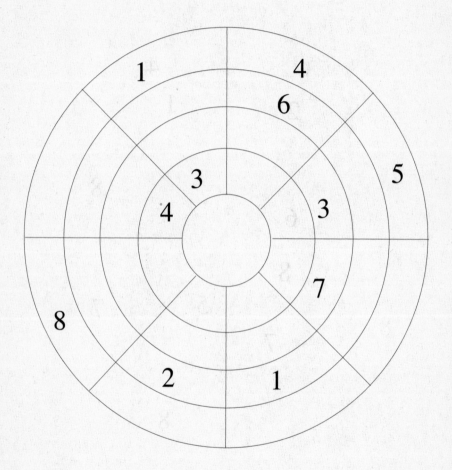

Puzzle 17 An empty middle ring.

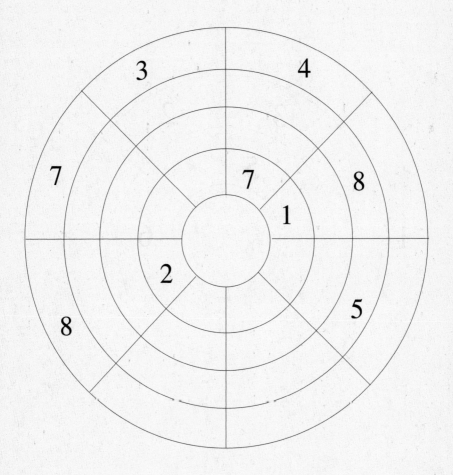

Puzzle 18   Another cute little three-ring puzzle!

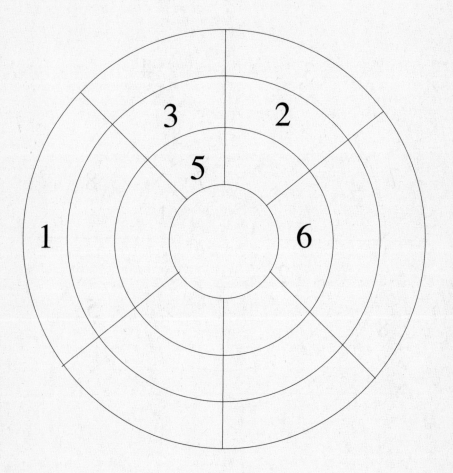

Puzzle 19 The clues of the next six spell out "Sudoku," beginning with "S." Use your imagination. . . .

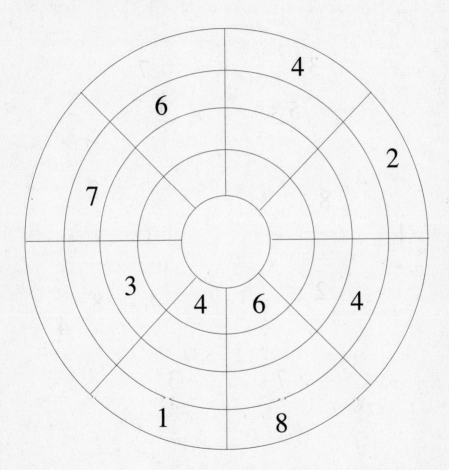

Puzzle 20 Spelling out the magic word, this time the clue is a kind of "u."

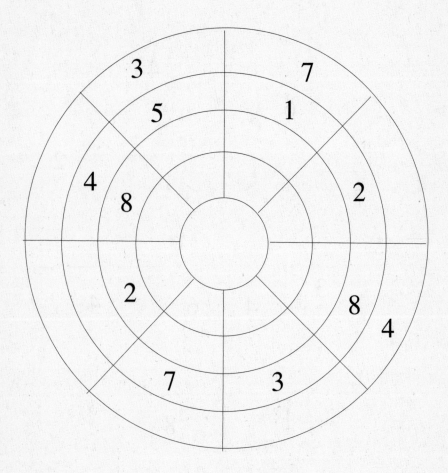

Puzzle 21 And next the "d."

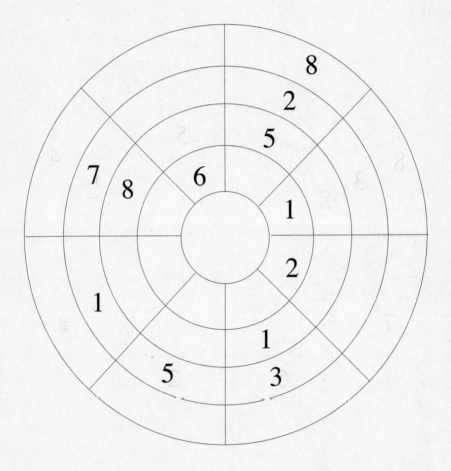

Puzzle 22 And now the "o."

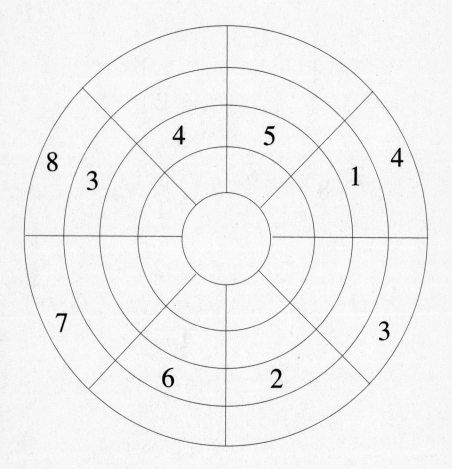

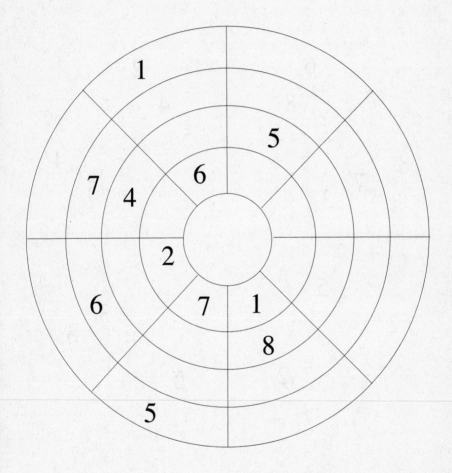

Puzzle 24 And the final "u."

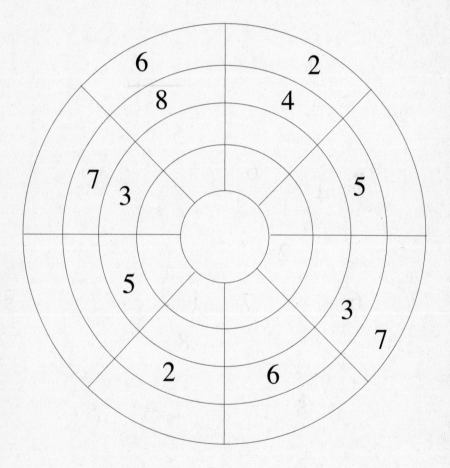

Puzzle 25 And, to complete this series, the first letter of the authors' last name.

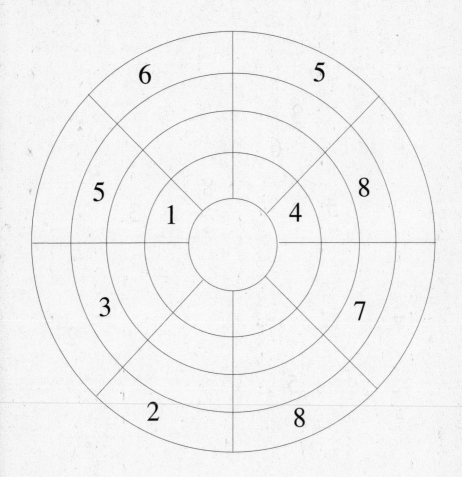

Puzzle 26   Another stingy example: just nine clues.

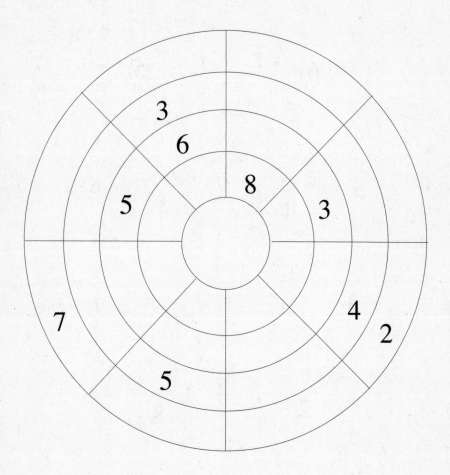

Puzzle 27 For once, no missing number!

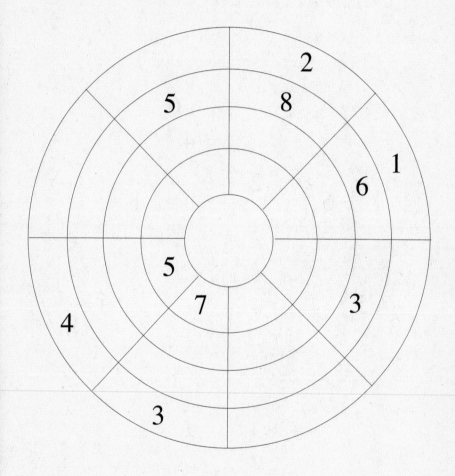

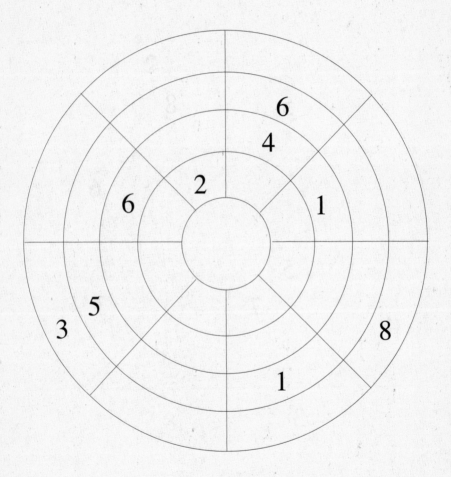

Puzzle 29 You shouldn't need any help with ten clues.

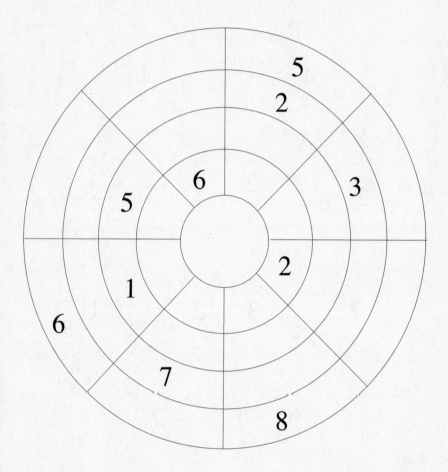

Puzzle 30   As in regular Sudoku, we can arrange the clues symmetrically.

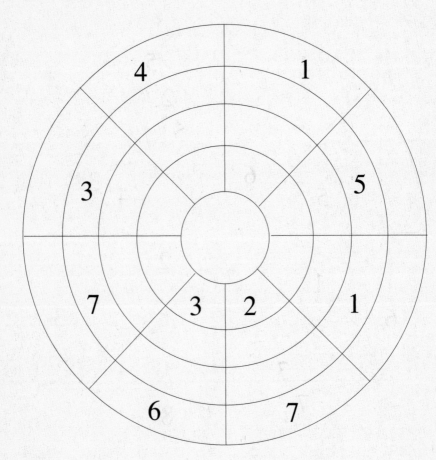

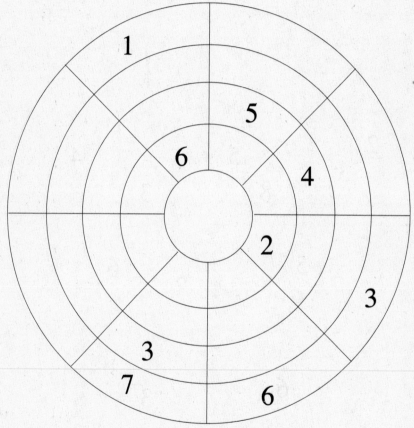

Puzzle 32 And four more to complete our first chapter of puzzles.

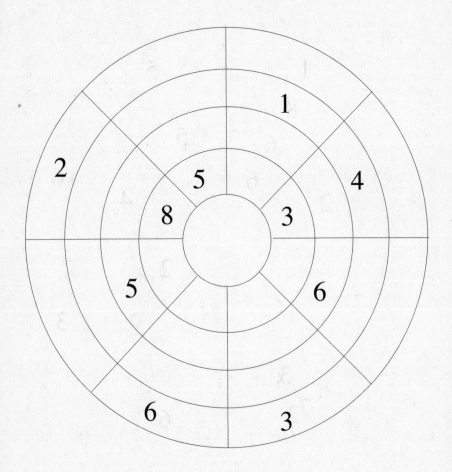

Puzzle 33   A little top-heavy.

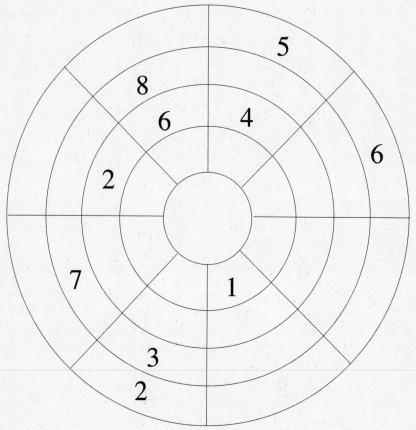

Puzzle 34 More clues in the southern hemisphere.

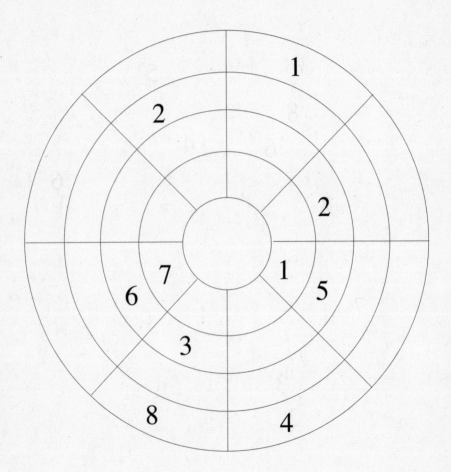

Puzzle 35   And finally the minimum nine.

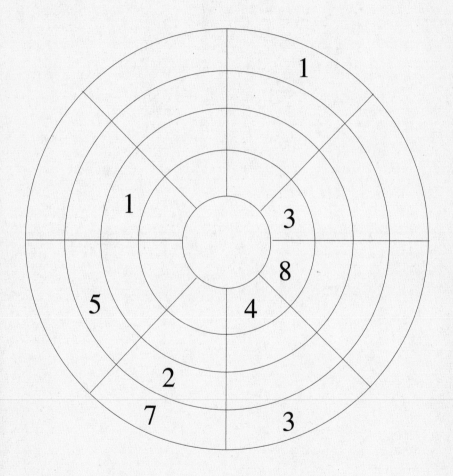

Five Rings
and Ten Symbols

The rule:

Five-ring puzzles: The numbers 1–9 are to appear in each of the five rings and each pair of adjacent slices of the circle.

Puzzle 36   Here's your first five-ring puzzle.

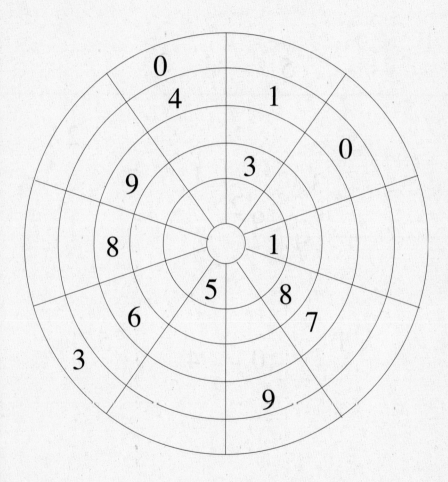

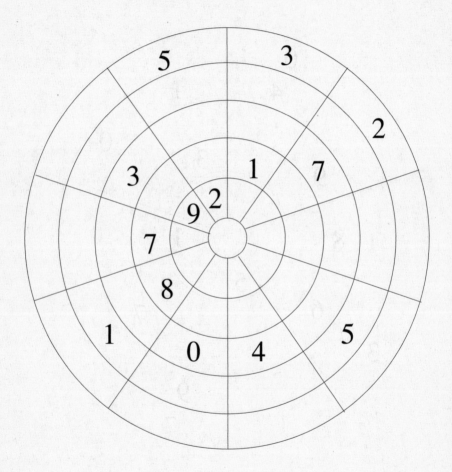

Puzzle 38   More clues don't always make for an easier puzzle!

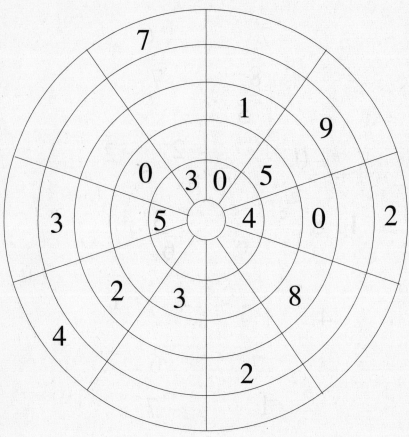

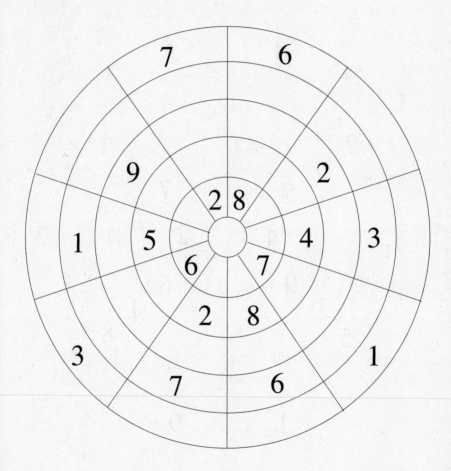

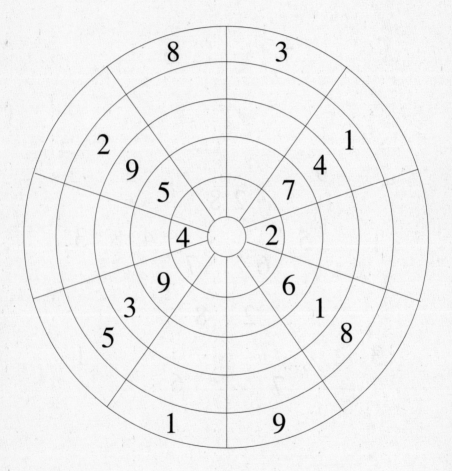

Puzzle 42   The next puzzle calls for a mixture of skills.

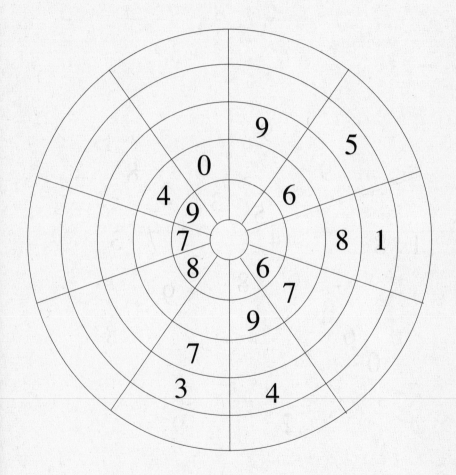

Puzzle 43 Every slice has a triple gap.

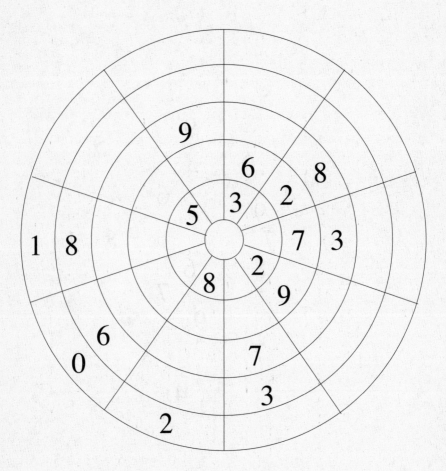

Puzzle 44   From now on, we'll mostly leave you to get on with it!

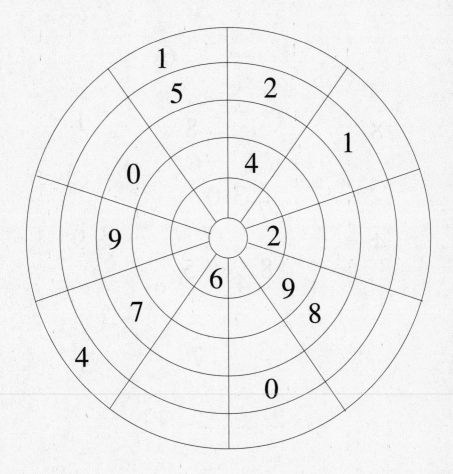

Puzzle 45   A crowded center.

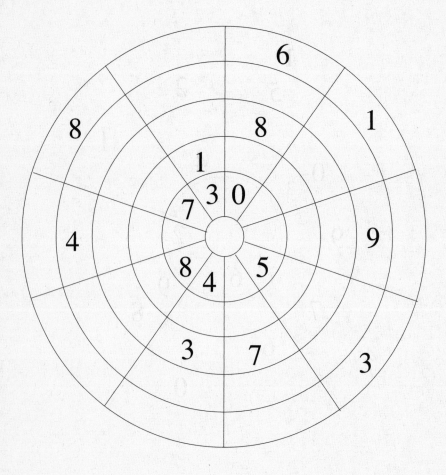

Puzzle 46 A low-high split.

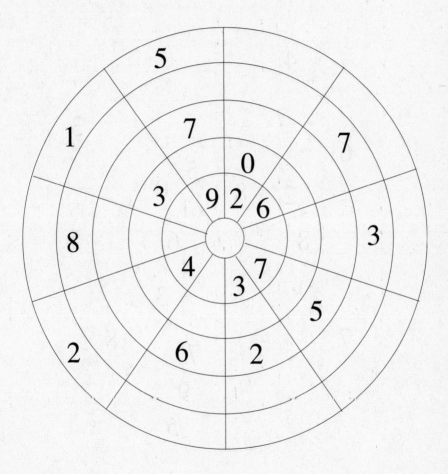

Puzzle 47 Big messy zigzag!

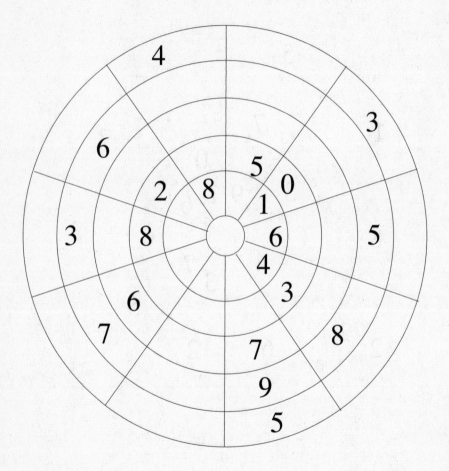

Puzzle 48 Little rhyme or reason.

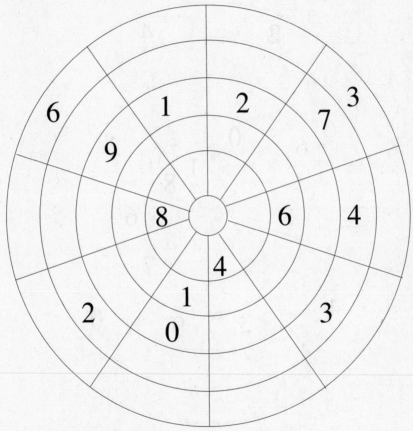

Puzzle 49 The 5's have gone missing.

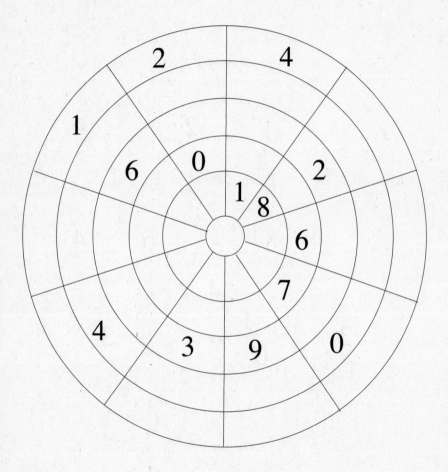

Puzzle 50 Here we have an empty ring and the fewest clues you will see—just thirteen in all.

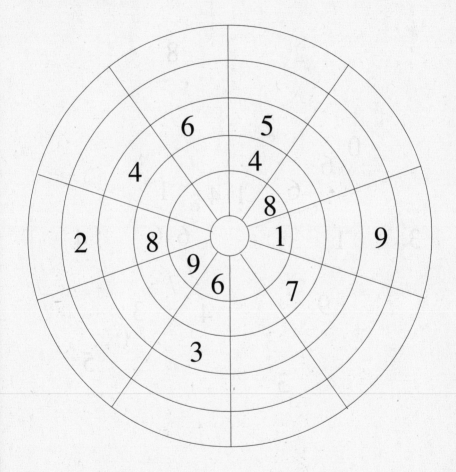

Puzzle 51 For once, no missing number.

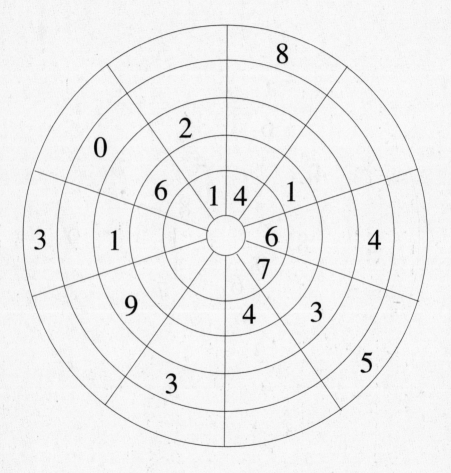

Puzzle 52 Does symmetry in the array make the puzzle easier?

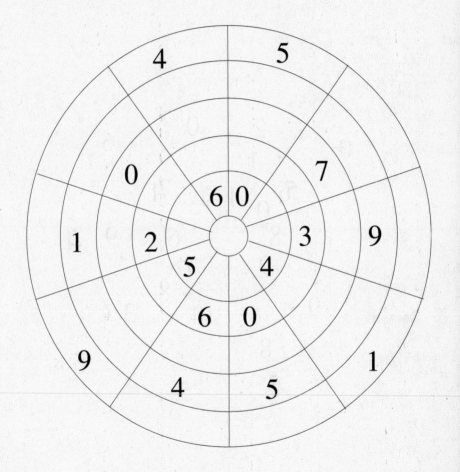

Puzzle 53 Nothing on the outside.

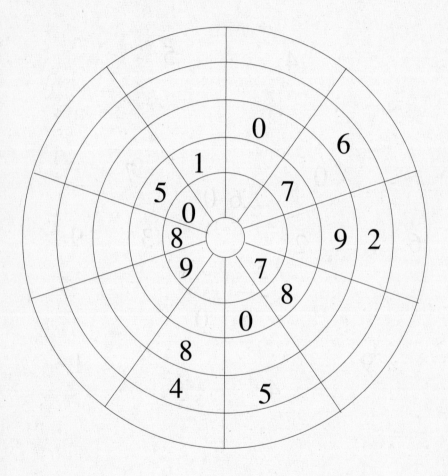

Symmetry of sorts: filled cells come in pairs and gaps in trios.

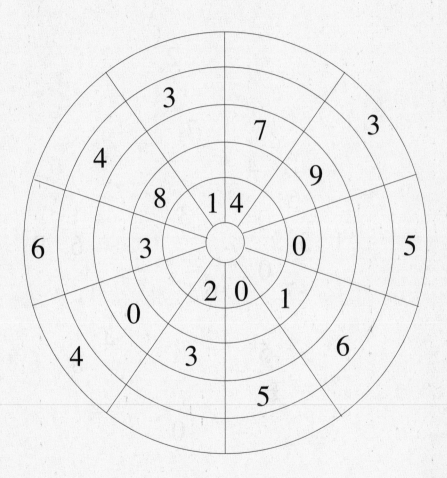

Puzzle 55 Crowded clues and open spaces.

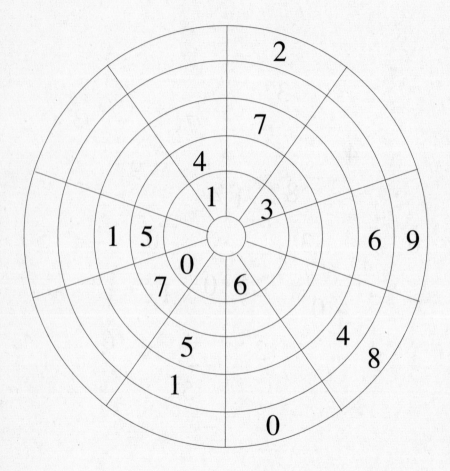

Puzzle 56 A random-looking puzzle.

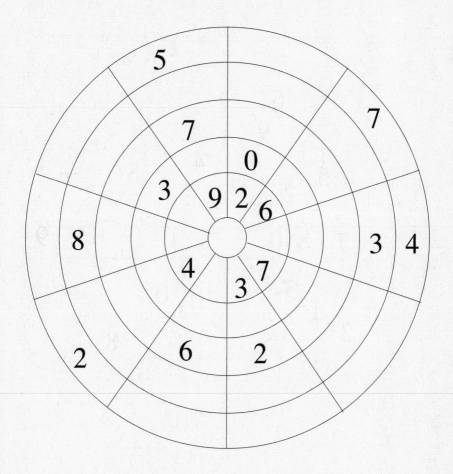

Puzzle 57 On this occasion you face the challenge of a pair of adjacent slices plus an entire ring bereft of clues.

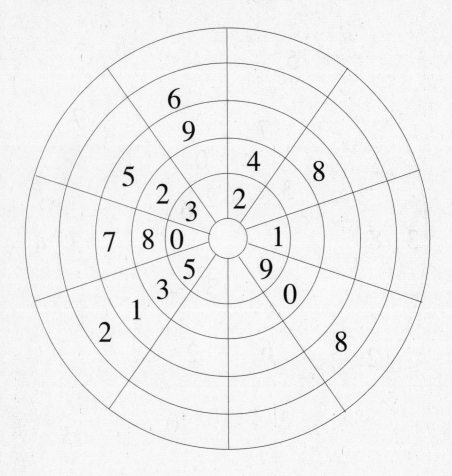

Puzzle 58 By way of contrast, the clues in this puzzle are spread about more evenly.

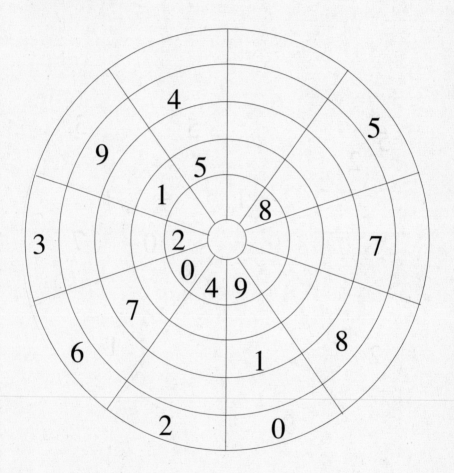

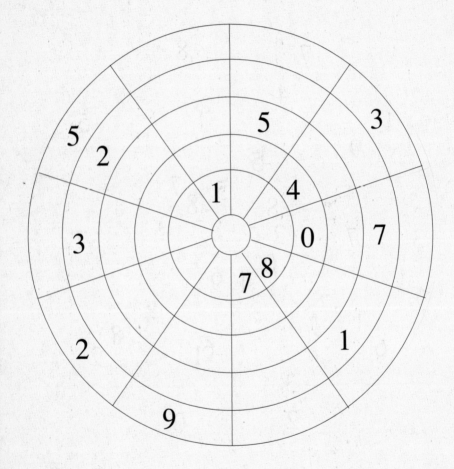

Puzzle 60   It might take a bit of thought to get started on this one, but then you should be away!

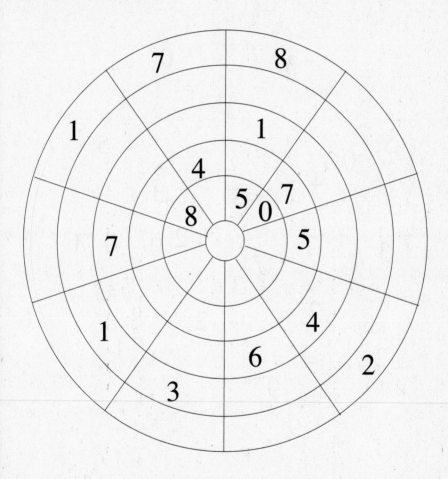

Puzzle 61 Again, you have at least one clue in every section.

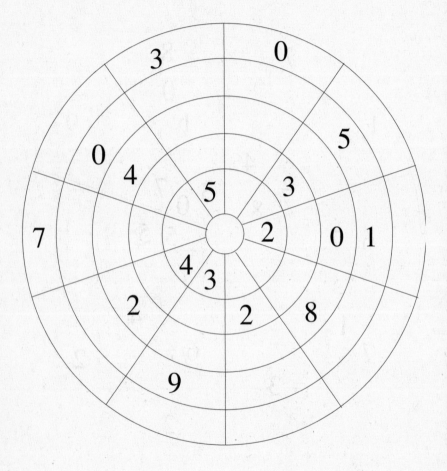

Puzzle 62   This time, every number is there to be seen.

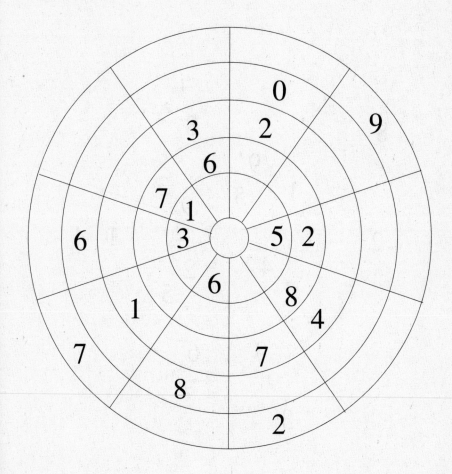

Puzzle 63 Can you tell if it is easy or not just by looking at it?

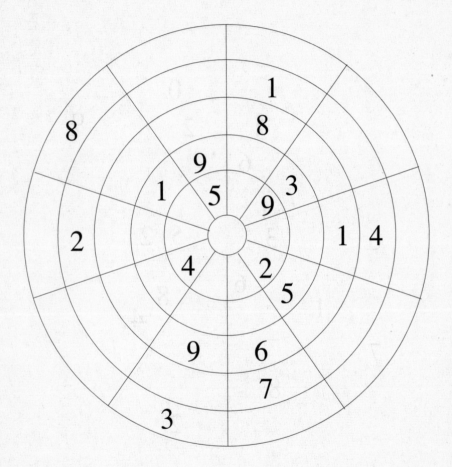

Puzzle 64 Again, a pair of vacant slices might make this puzzle a challenge!

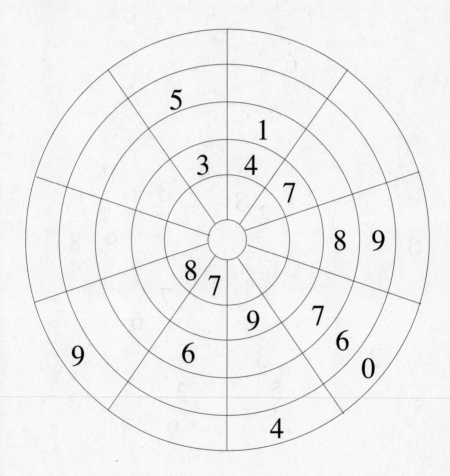

Puzzle 65 A lot of clues should make it easier?

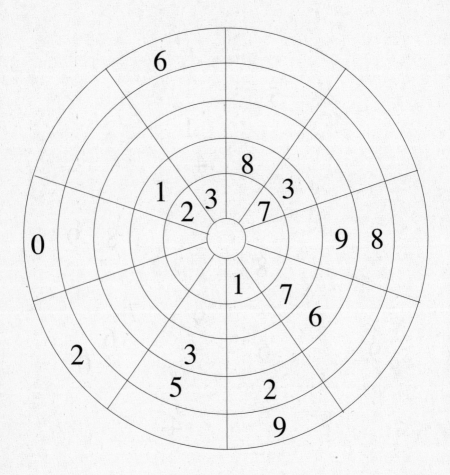

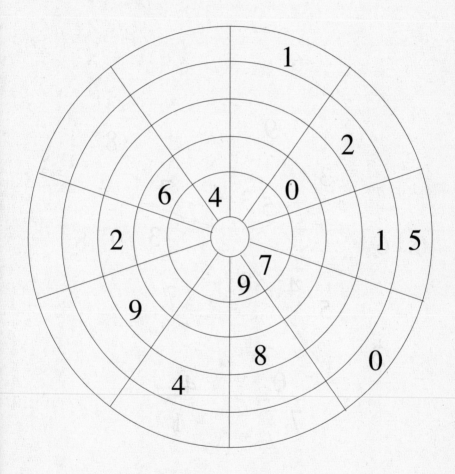

Puzzle 67 The clues are more on the outer.

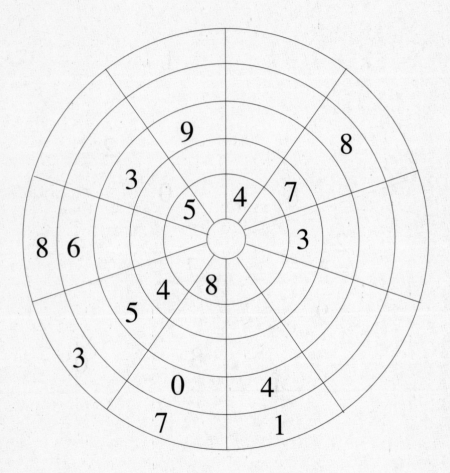

Puzzle 68   More clues than you often need.

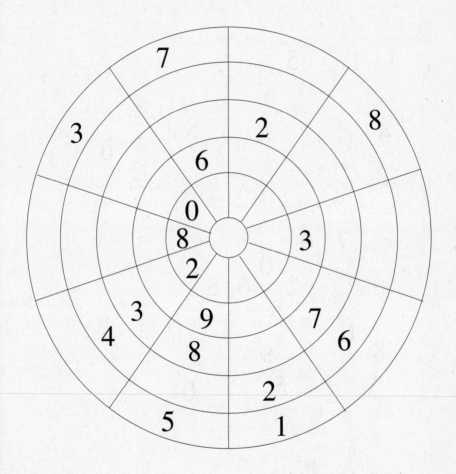

Puzzle 69 Here you are nearly given a whole slice.

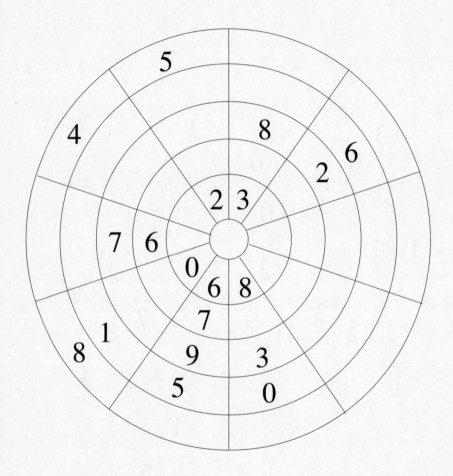

Puzzle 70   Back to just fourteen given cells.

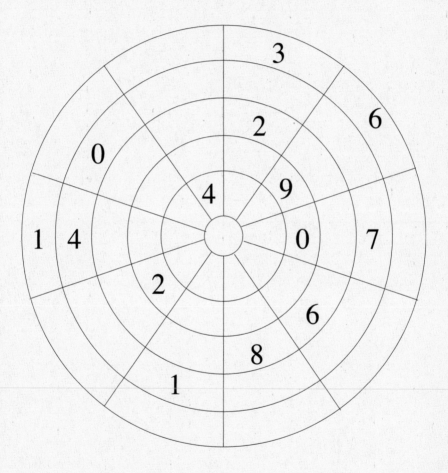

Variations on a Theme

Ring Sudoku
Double Sudoku
Target Sudoku

. . . and other brain-melting puzzles for experts

Puzzle 71 Magic Circle Sudoku! Every ring sums to twice every slice!

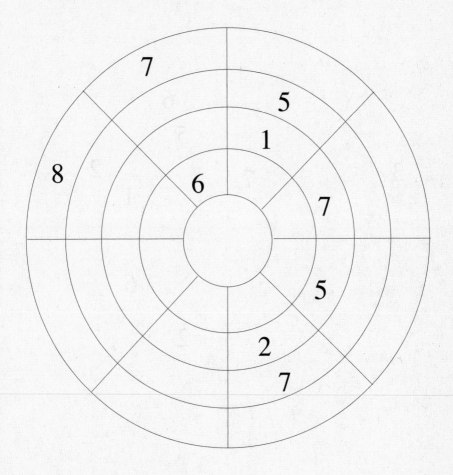

Puzzle 72   Only eight clues and two missing numbers, which is normally illegal—but once again, it's magic!

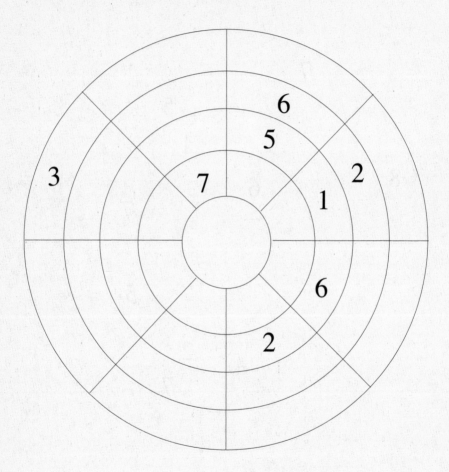

Puzzle 73 Only five clues—but there is another: consecutive numbers must not be neighbors, either within a ring, or a slice.

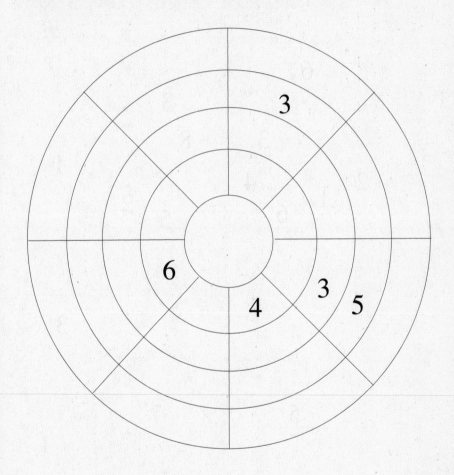

Puzzle 74   This time we have changed the rules: each number 1-8 has to appear in every ring and every pair of *diametrically opposite slices*.

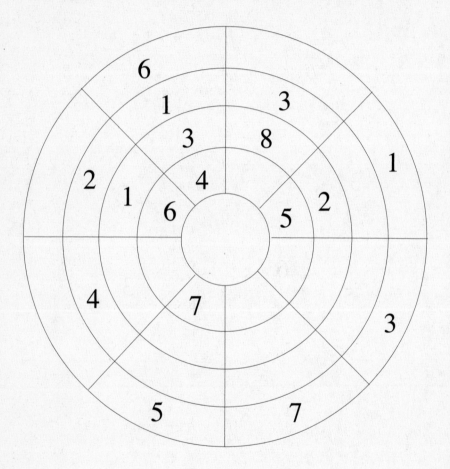

Puzzle 75 And once again, it is the opposite slice rule that applies.

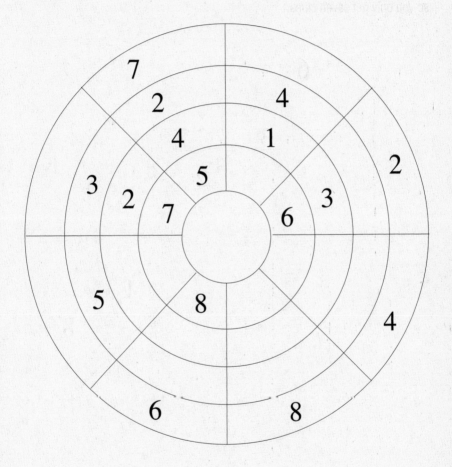

Puzzle 76   Ring Sudoku. The usual rules of the game apply but this time every ring is the same as every other, just turned around. It must be easy, so you only get seven clues.

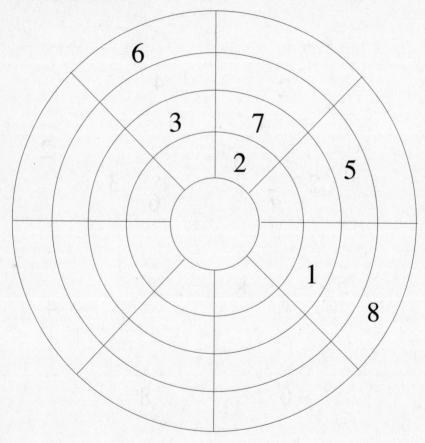

Again, it is a Ring Sudoku, so you have a safe to crack!

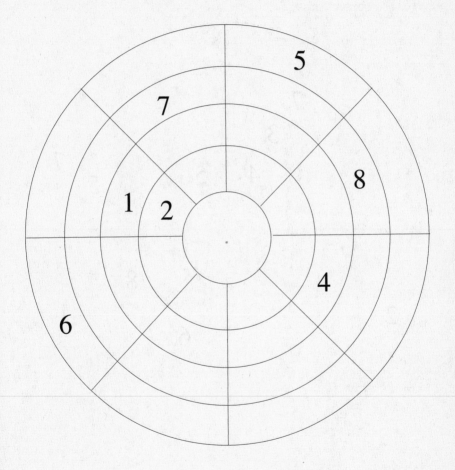

Puzzle 78 Once more, we have a Ring Sudoku but, to make things a little tougher perhaps, it's the opposite slice rule that applies.

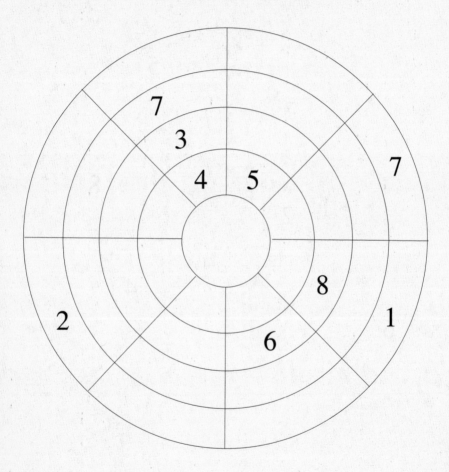

Puzzle 79   An opposite slice Ring Sudoku that spells out something a
little unusual! (Identifying the missing letter is part of the conundrum!)

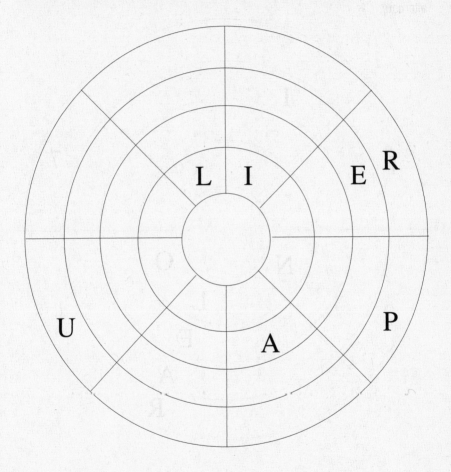

Puzzle 80   And finally in this series, an opposite slice Ring Sudoku
with a twist: each ring spells out a dedication, and vowels and consonants
alternate.

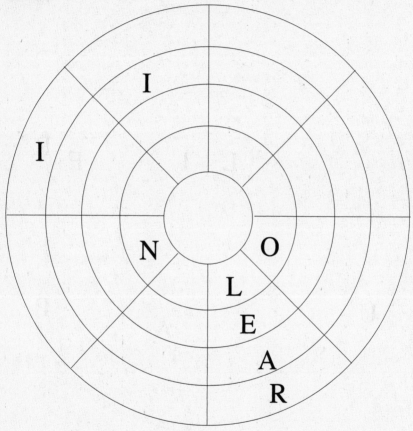

Puzzle 81   Only twelve clues. You will need another: extremes must be kept apart.

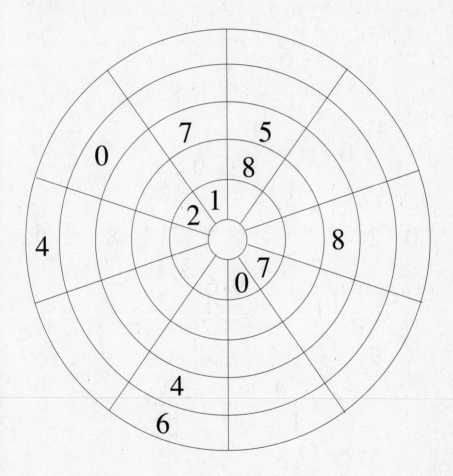

Puzzle 82 A five-ring opposite slice puzzle—lots of clues so it must be easy, musn't it?

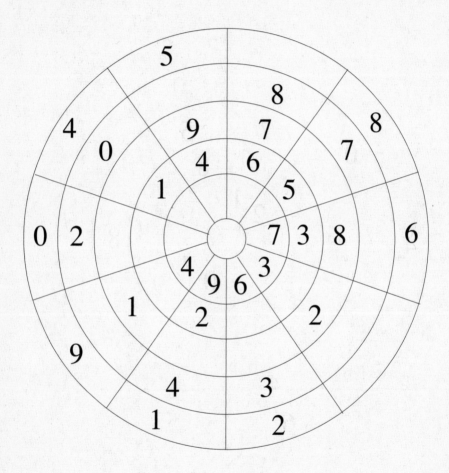

Puzzle 83 And another.

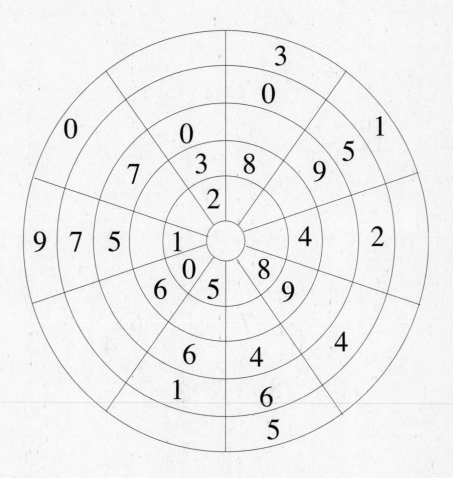

Puzzle 84   Like Puzzles 76 and 77, this is a Ring Sudoku—all the rings are the same, but otherwise the rules are as normal.

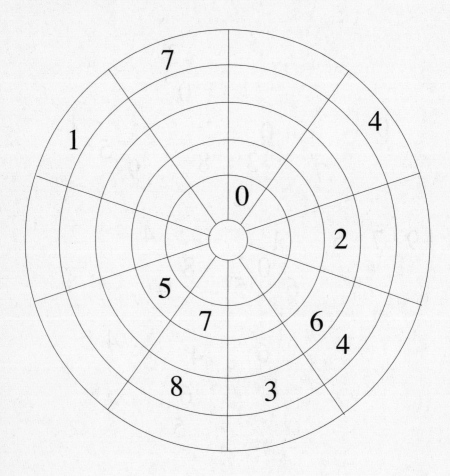

Puzzle 85 Ring Sudoku, but this one is a little more challenging.

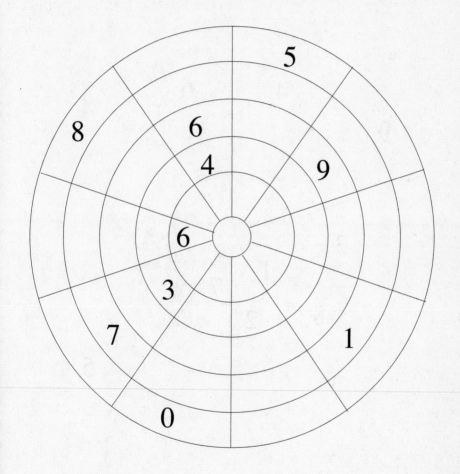

Puzzle 86 Were the last pair of Ring Sudoku too easy? Then try this one, with the minimum possible number of clues!

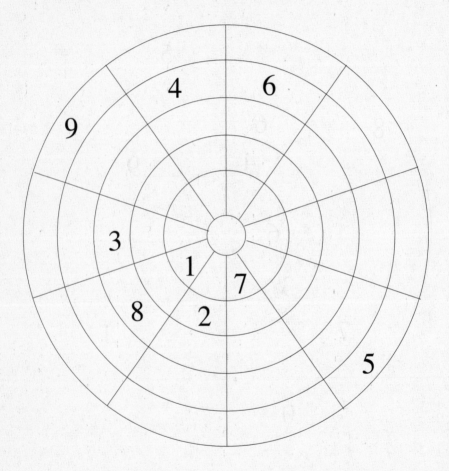

Puzzle 87   This is the toughest one—Ring Sudoku with the opposite slices rule.

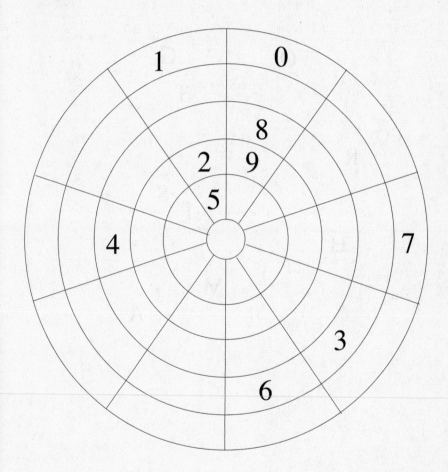

Puzzle 88   Adjacent slices Ring Sudoku—this one has a
mathematical bent.

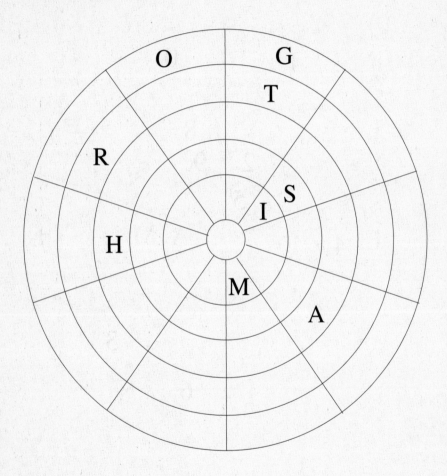

Puzzle 89 This standard Ring Sudoku is not to be trusted!

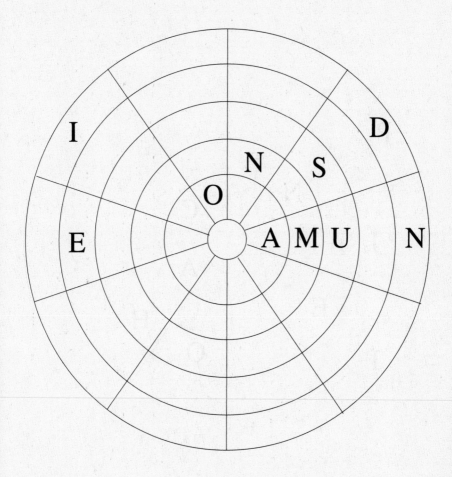

Puzzle 90 To make it a little harder, this time the mystery word is embedded into an opposite slice Ring Sudoku.

Puzzle 91 And next we graduate to six-ring regular Circular Sudoku—the twelve symbols now include A and B.

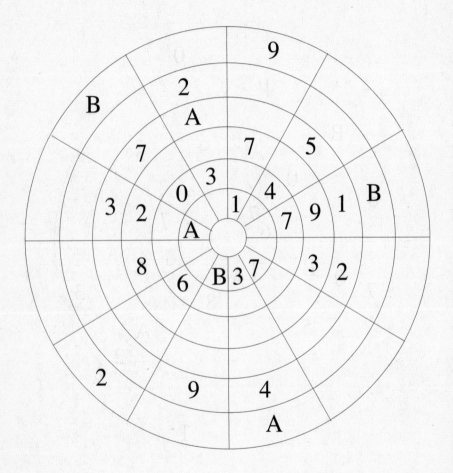

Puzzle 93 And a third.

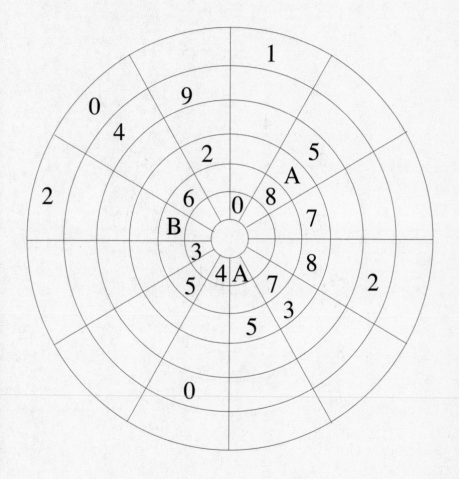

Puzzle 94 This time you have another wall safe to open—it is a Ring Sudoku (with the normal adjacent slice rule).

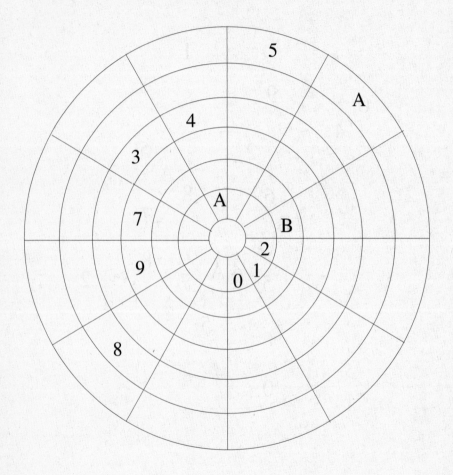

Puzzle 95   Like Puzzle 89, we have a word Ring Sudoku with the standard rules but with six rings (and twelve letters).

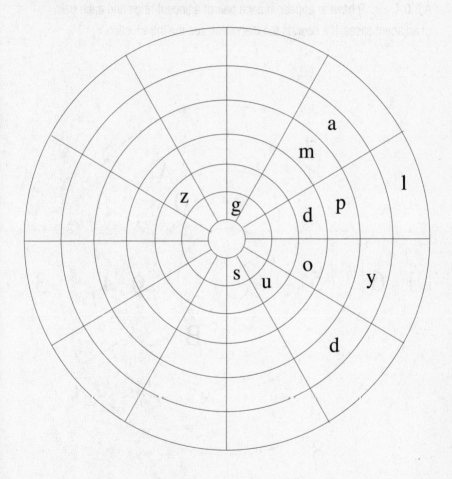

Puzzle 96   The first of five Double Sudoku puzzles. Here the same rule applies to both rings and slices—each of the twelve symbols A,B,0,1, . . . ,9 have to appear in each pair of adjacent rings and each pair of adjacent slices. If it sounds too diabolical, see the Introduction.

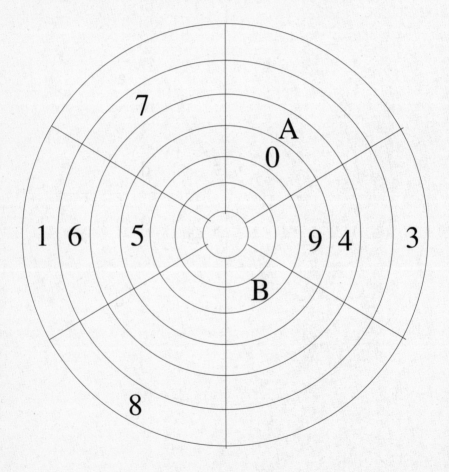

Puzzle 97 See if you can manage this Double Sudoku in less time.

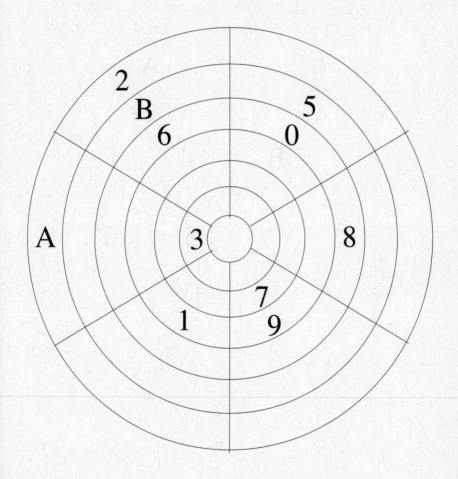

Puzzle 98 Does symmetry make Double Sudoku easier?

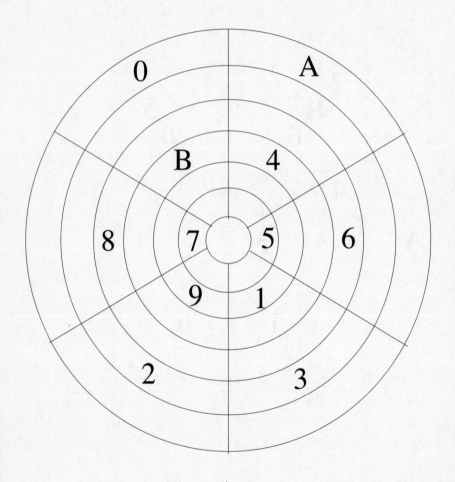

Puzzle 99 Again, we offer a left-right mirror image in the pattern of occupied cells.

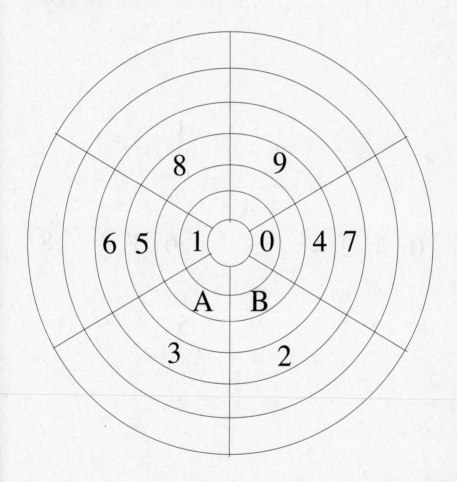

Puzzle 100 And finally in this series, near symmetry with a missing symbol.

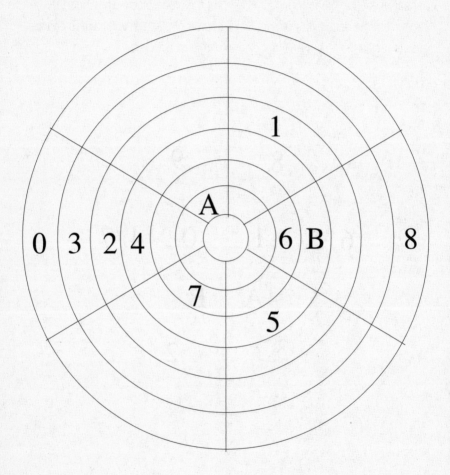

Puzzle 101 Target Sudoku, the first of a whole new type. The rule for the WBW (white-black-white) puzzle is that each of the twelve symbols must appear in each of the four rings and each of the white-black-white quarter circles (but not necessarily in each BWB quarter circle). You will have to change gear to tackle these, but this first example at least is gentle.

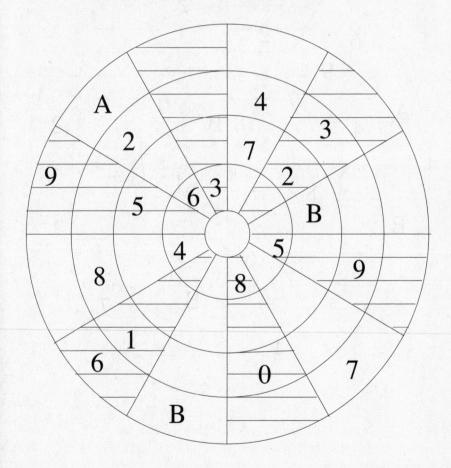

Puzzle 102 Another gentle one to get you into the swing.

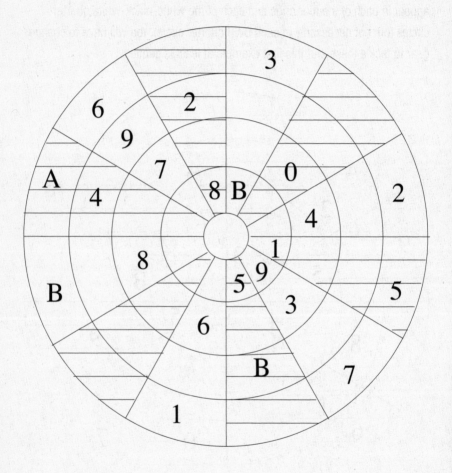

Puzzle 103 Lots of empty space in this one—an empty slice and only nineteen clues with no number appearing more than twice. A little more difficult?

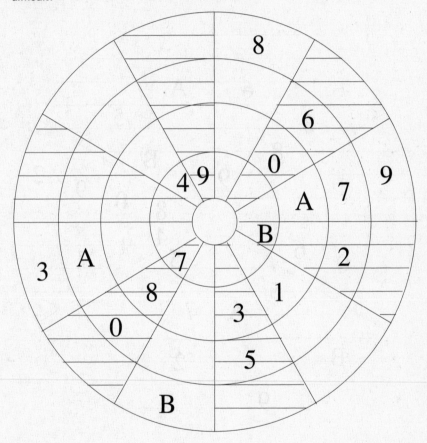

Puzzle 104   Another with plenty of clues.

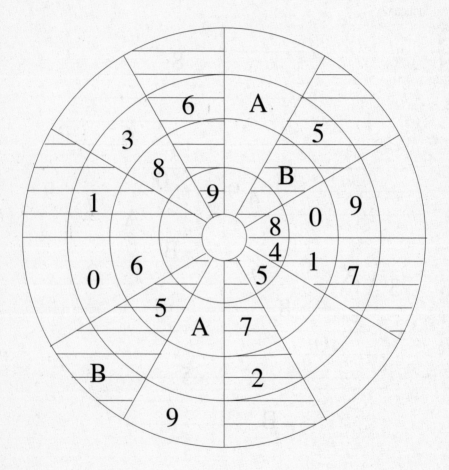

Puzzle 105 The filled cells are more evenly spread.

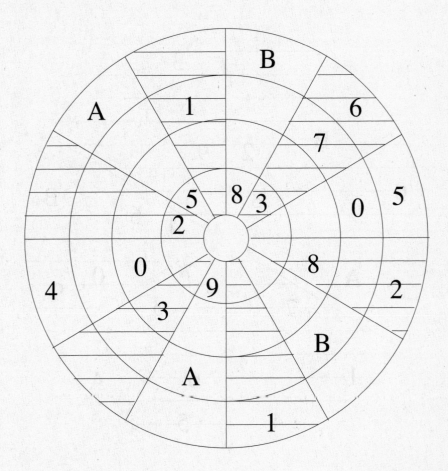

Puzzle 106   Again, this is fairly gentle, although no number appears more than twice as a clue.

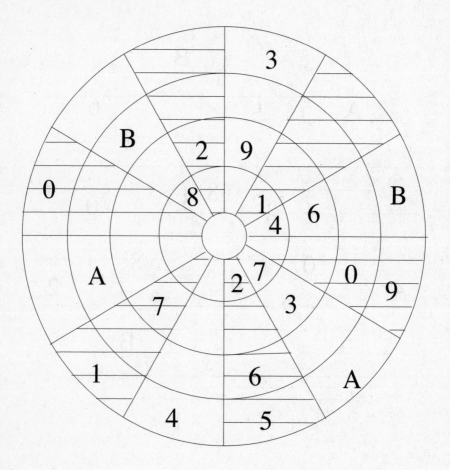

Puzzle 107 This one is a bit sneaky; see if you can work out how to solve it.

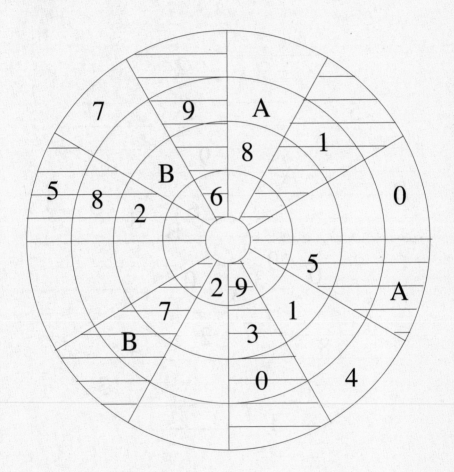

Puzzle 108 Half the array is given away, but does that make it so easy?

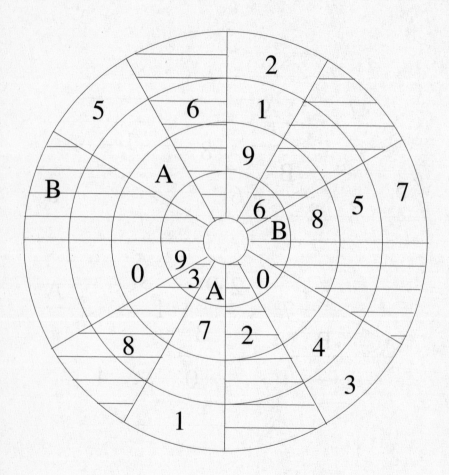

Puzzle 109   For the first time, we find one number is missing entirely from the given scheme.

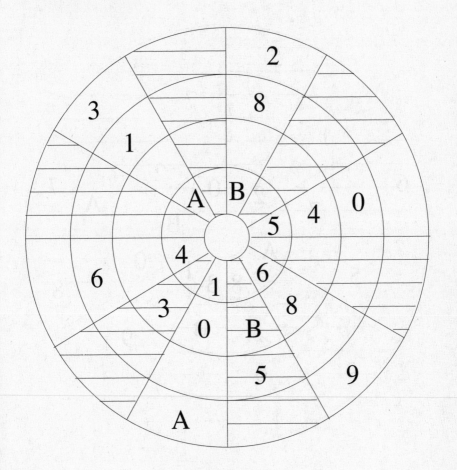

Puzzle 110 Again, the level of difficulty is a little higher than some of the earlier Target puzzles.

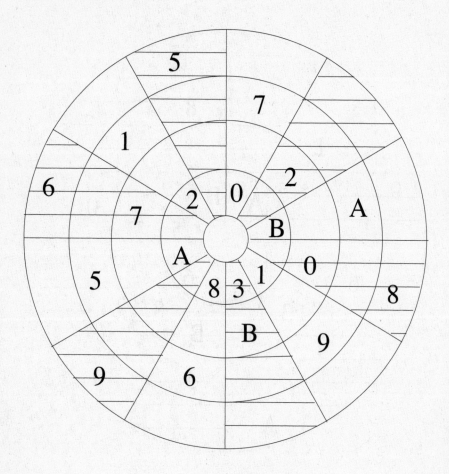

Puzzle 111 Are the puzzles getting easier or are you getting more clever?

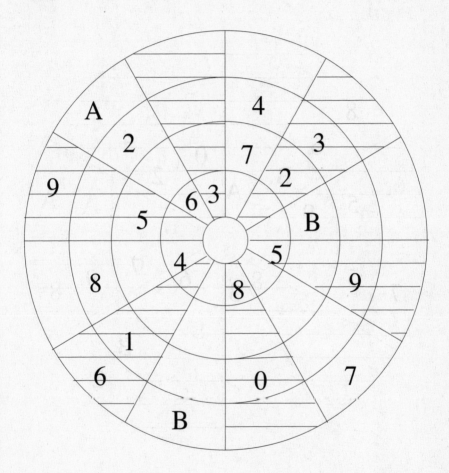

Puzzle 112 The difficulty is stepped up a level with this one.

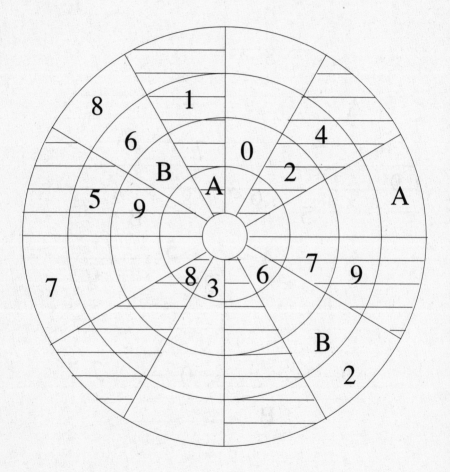

Sometimes you have to keep in mind the "big picture."

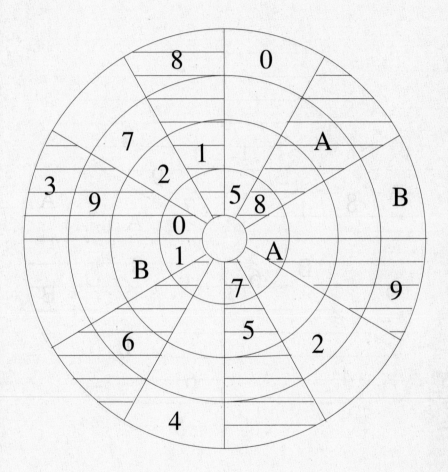

Puzzle 114 Once again, one number has gone AWOL!

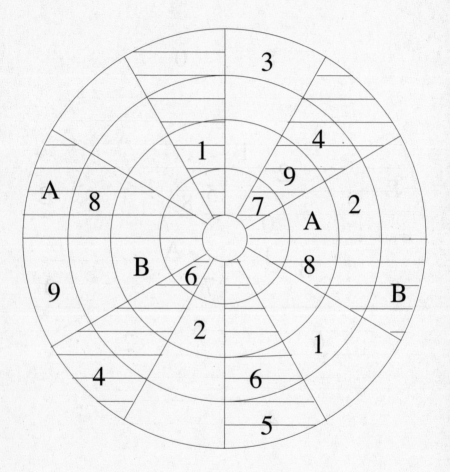

Puzzle 115   Again, just twenty clues and 1 is nowhere to be seen as yet.

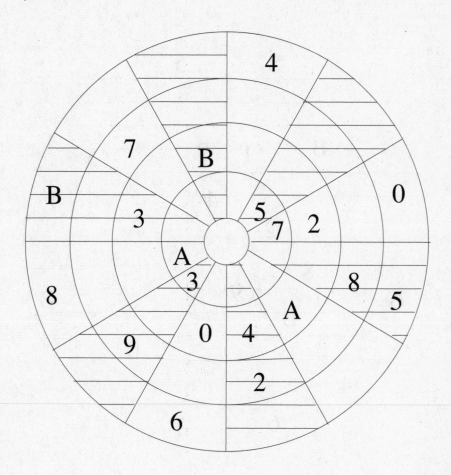

Puzzle 116 Another five-star Target.

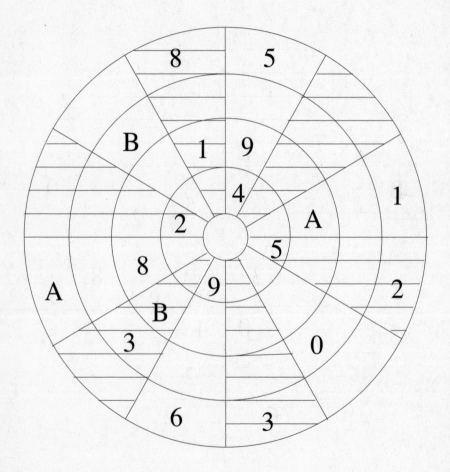

Puzzle 117

At least this time all the symbols are present if not yet all accounted for.

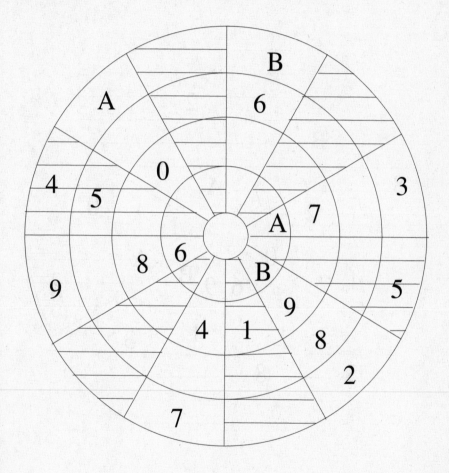

Puzzle 118 Once more a missing symbol makes this one a bit trickier.

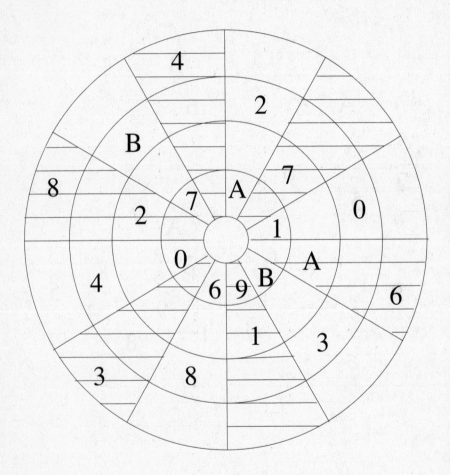

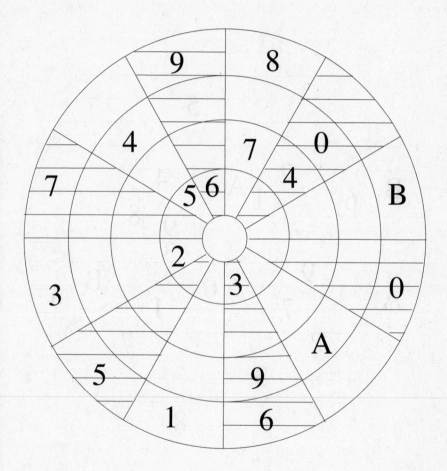

Puzzle 120   And for our final puzzle, things go full circle.

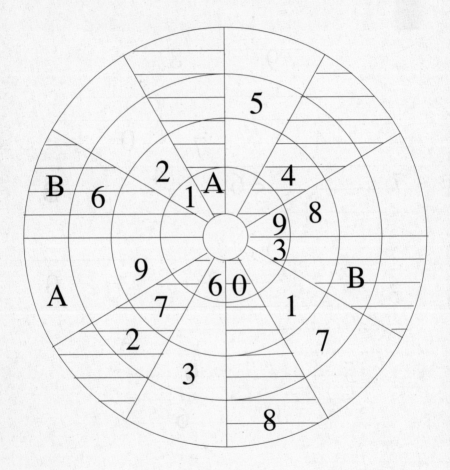

Solutions

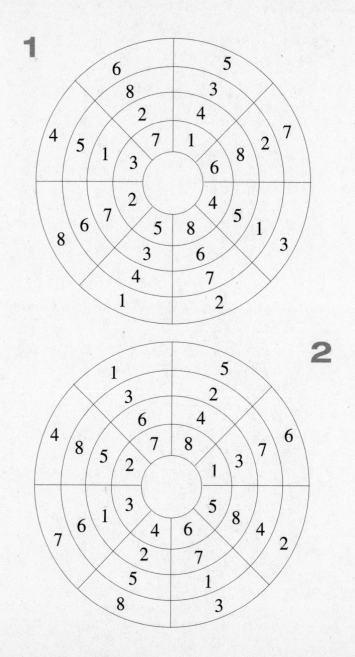

3

4

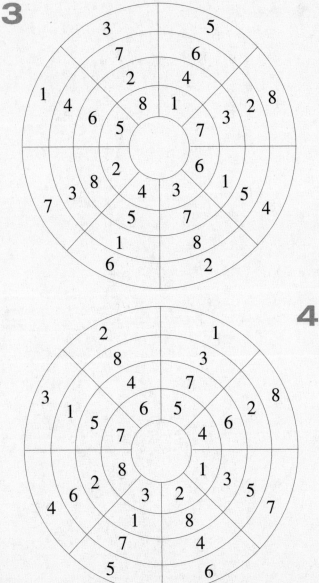

5

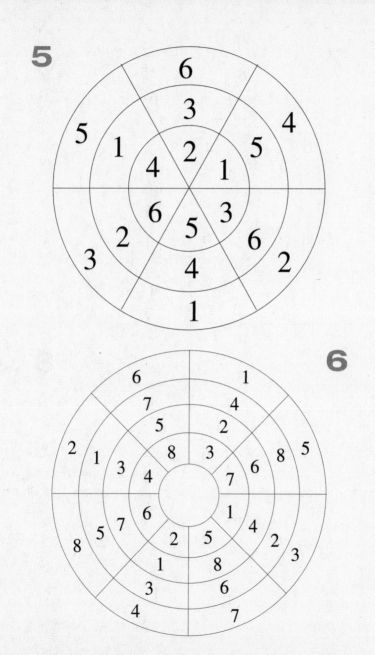

6

7

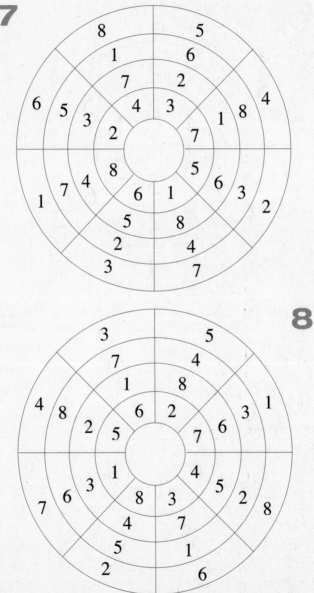

8

9

10

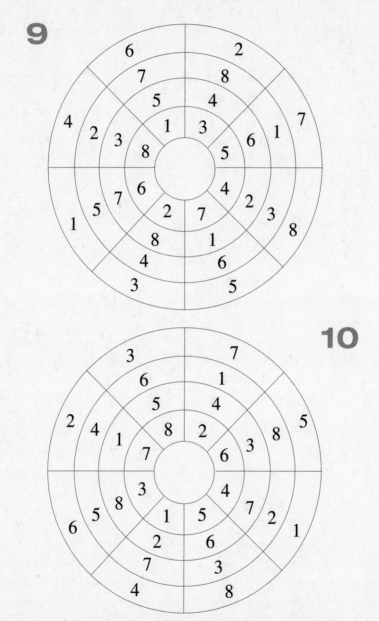

11

12

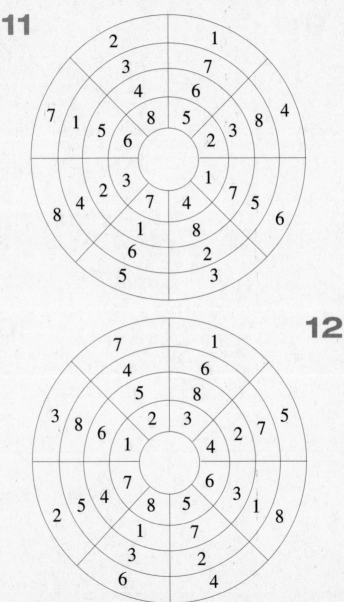

13

14

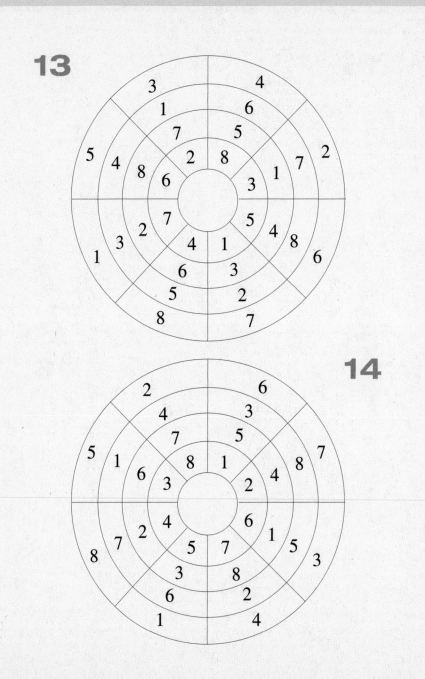

15

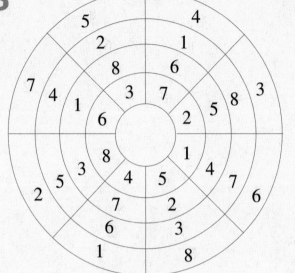

16

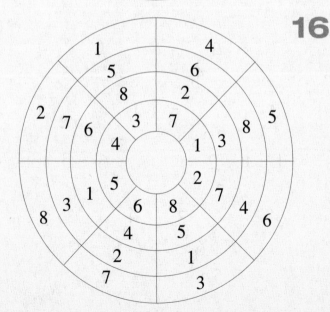

17

18

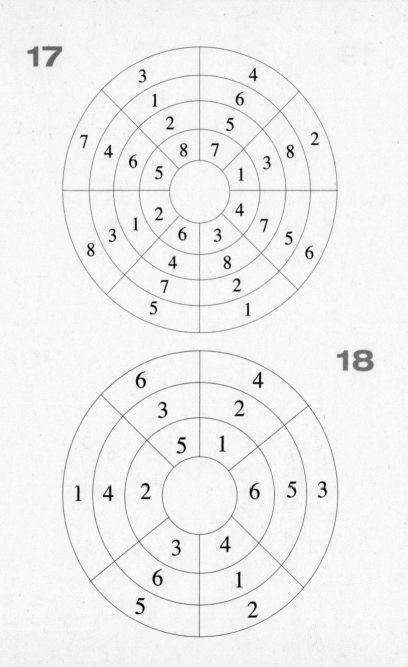

19

20

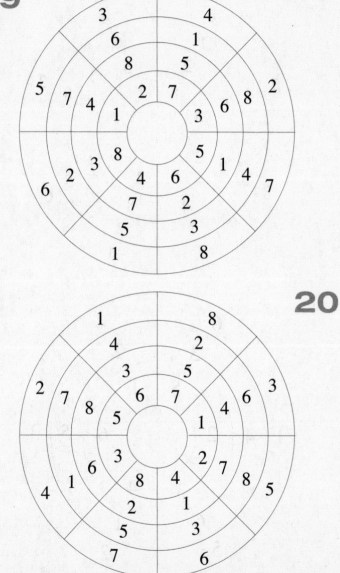

21

22

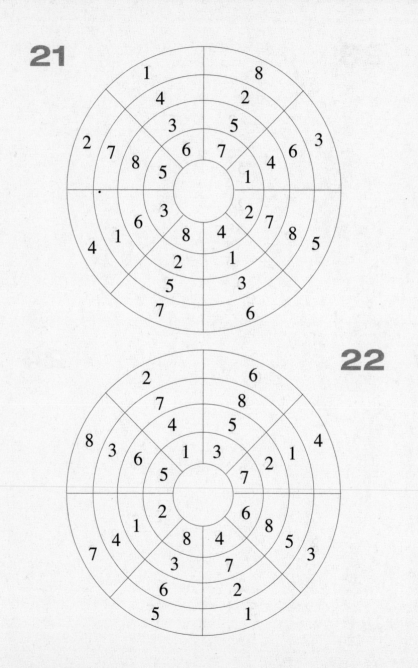

23

24

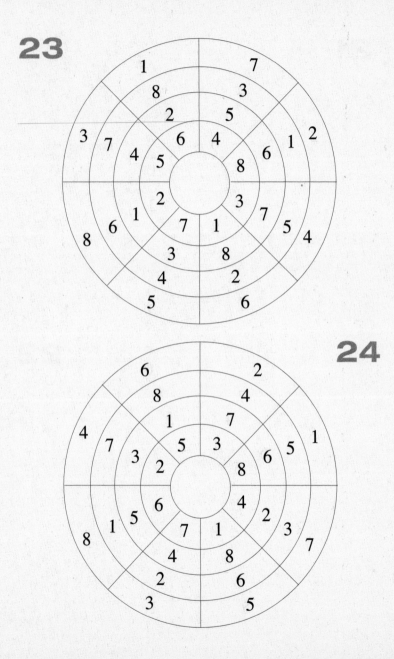

25

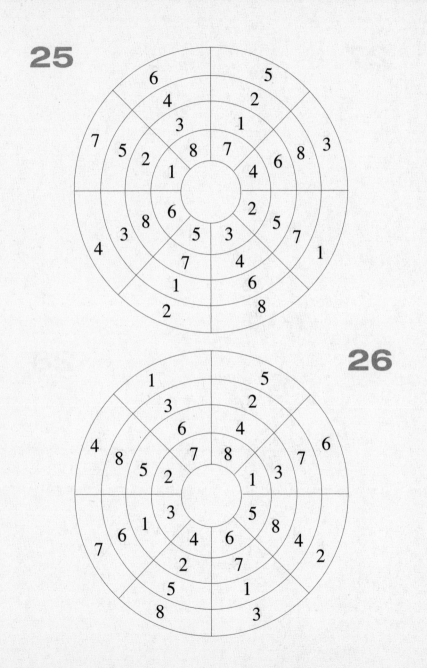

26

27

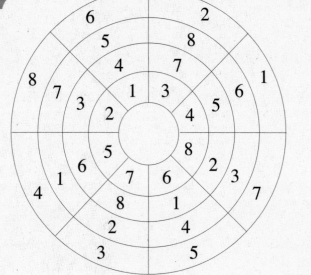

28

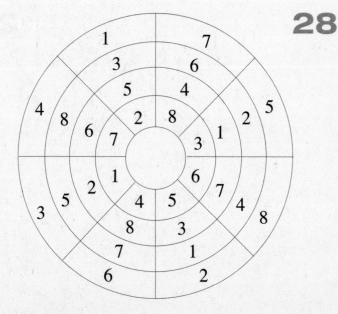

29

30

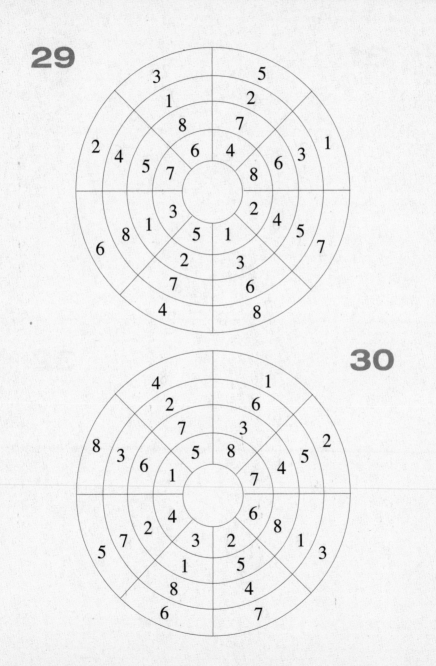

31

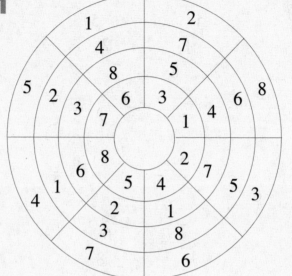

32

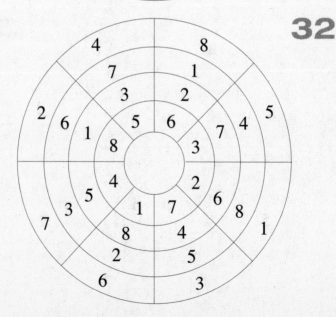

33

34

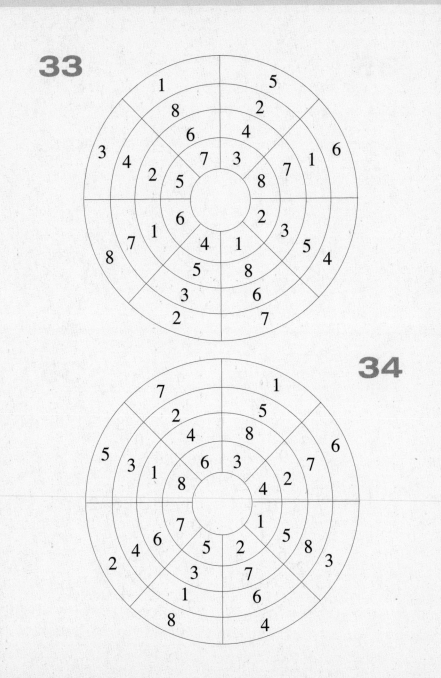

35

36

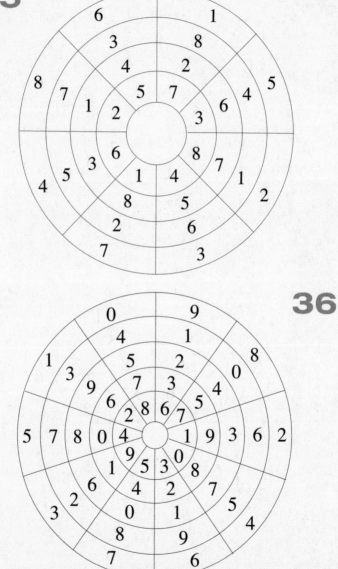

37

38

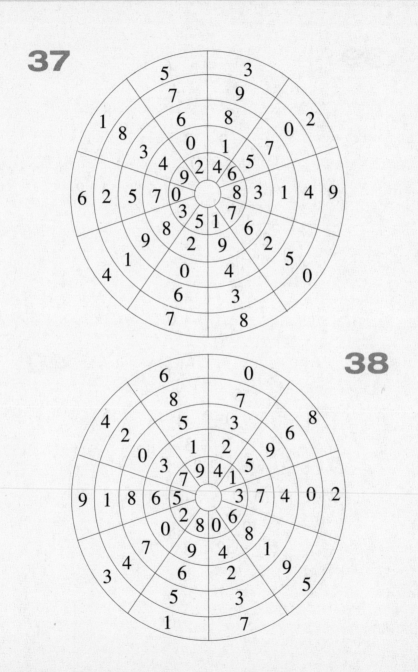

39

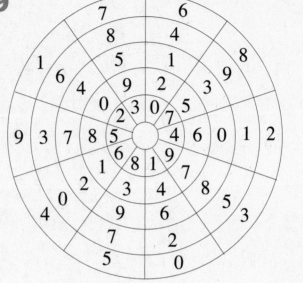

40

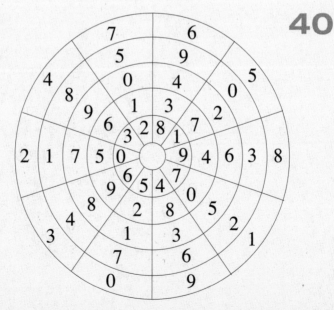

41

42

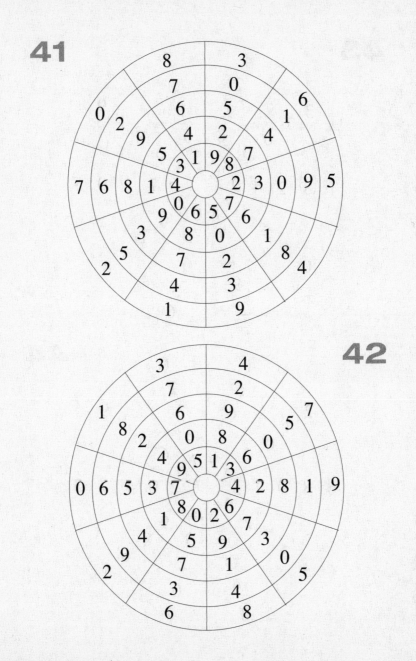

43

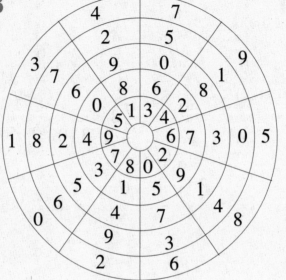

44

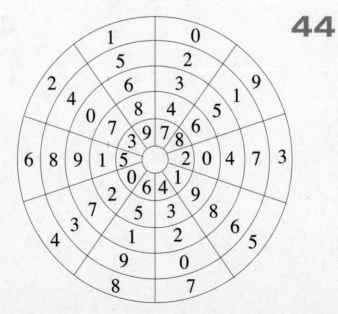

45

46

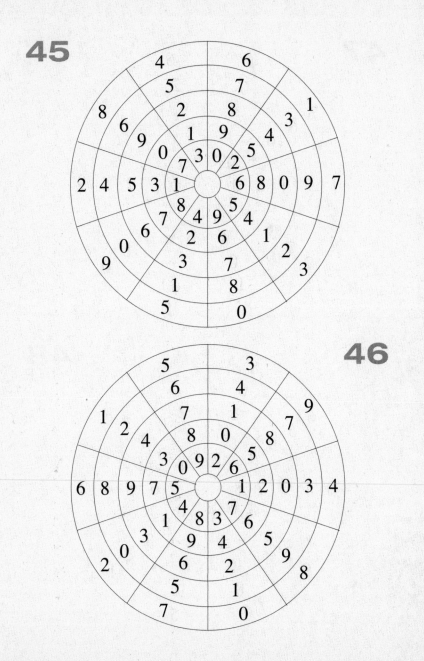

47

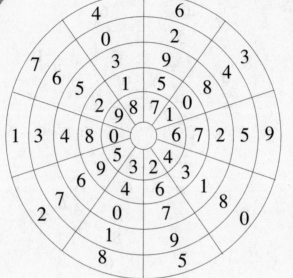

48

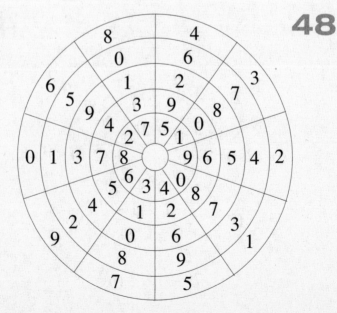

49

50

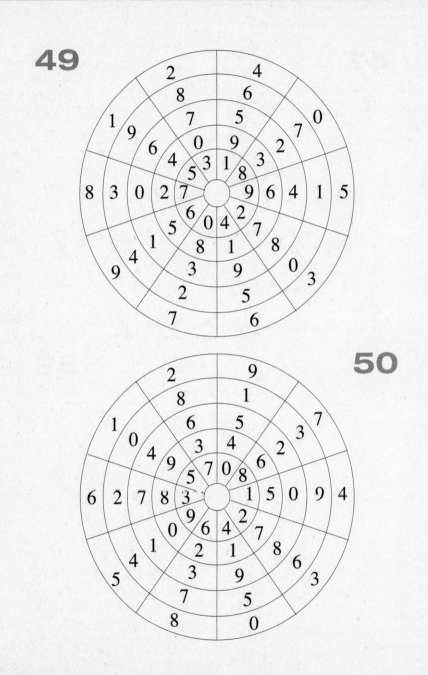

51

52

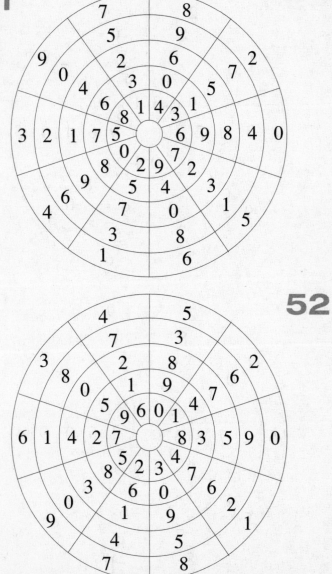

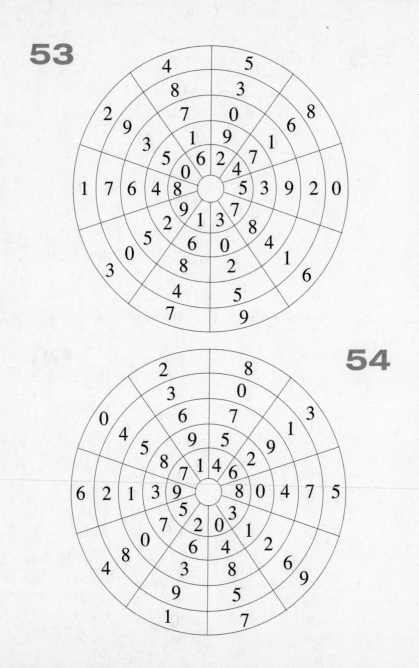

53

54

55

56

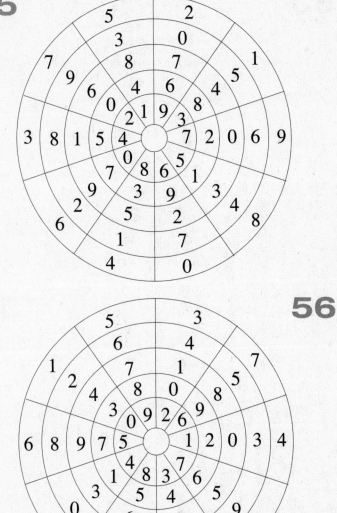

57

58

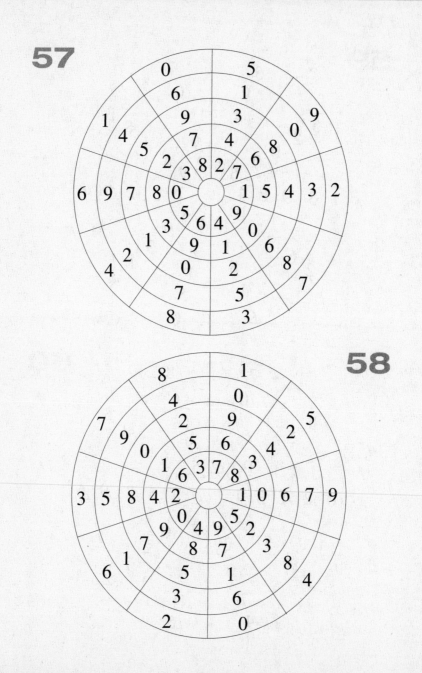

59

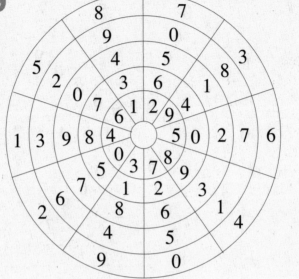

60

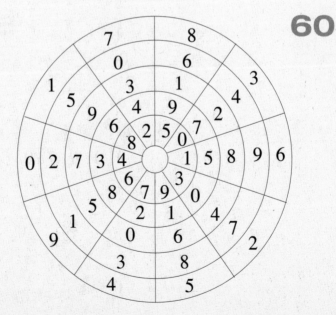

61

62

63

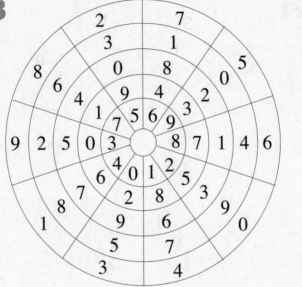

64

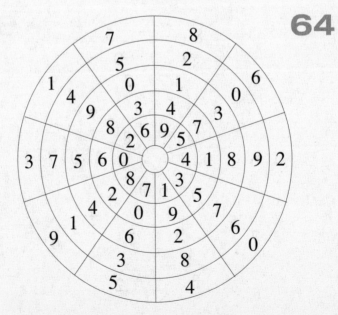

65

66

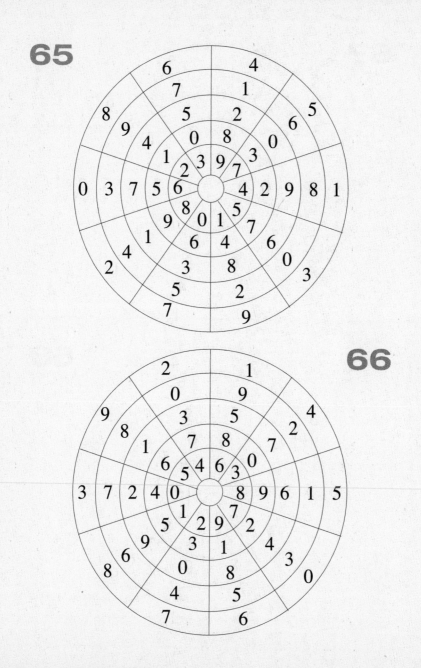

67

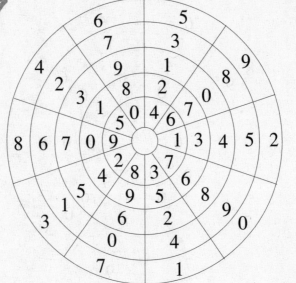

68

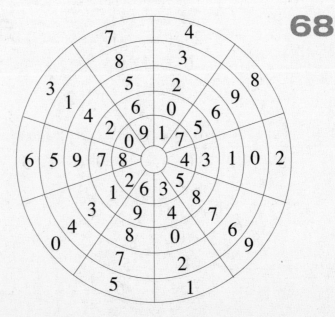

69

70

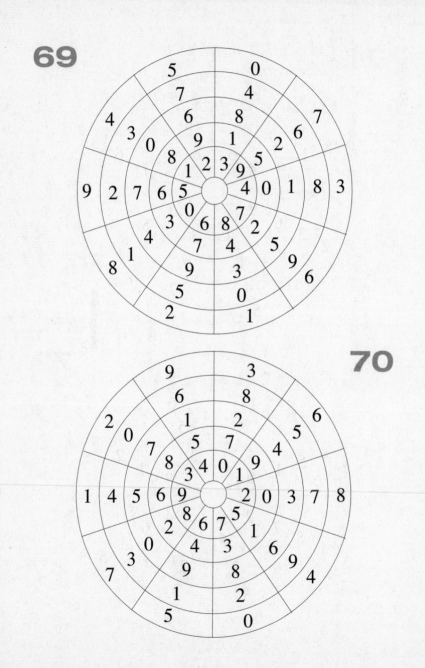

71

72

73

74

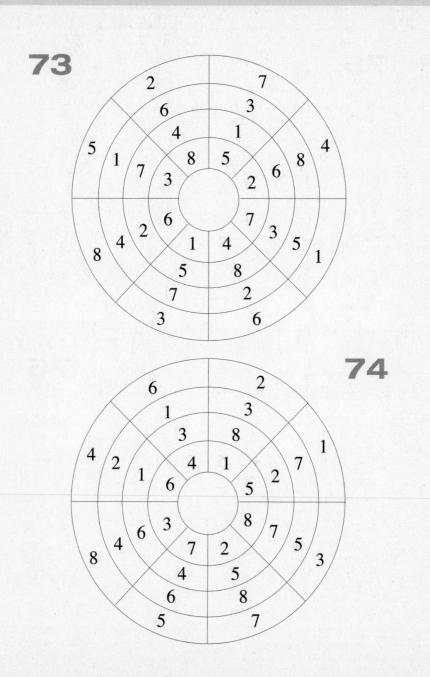

75

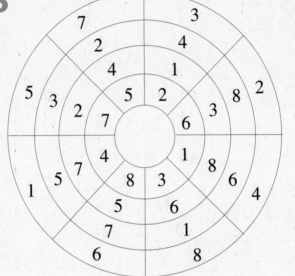

76

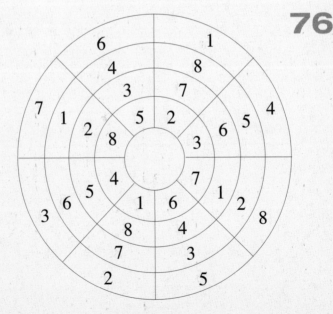

77

78

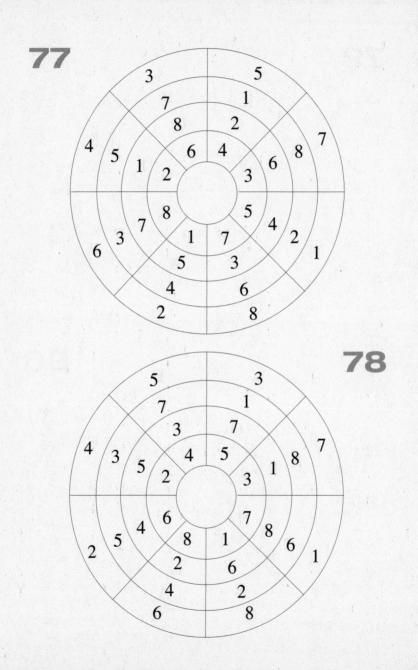

79

80

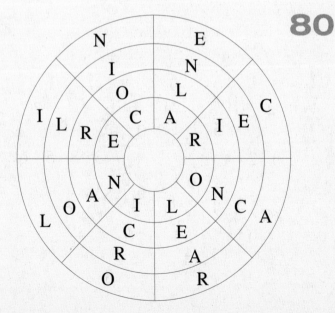

81

82

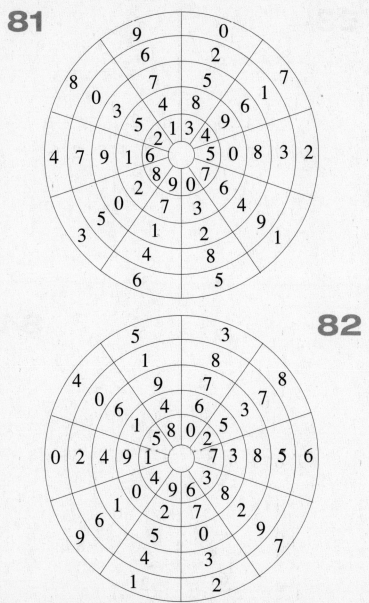

83

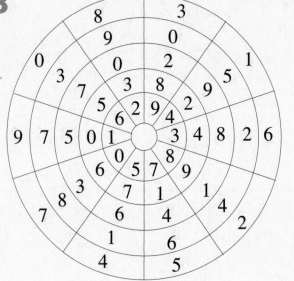

84

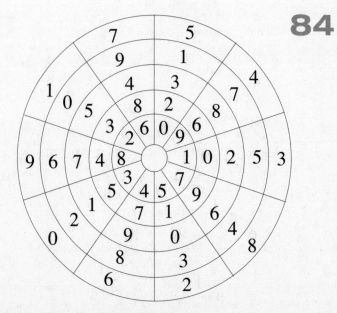

85

86

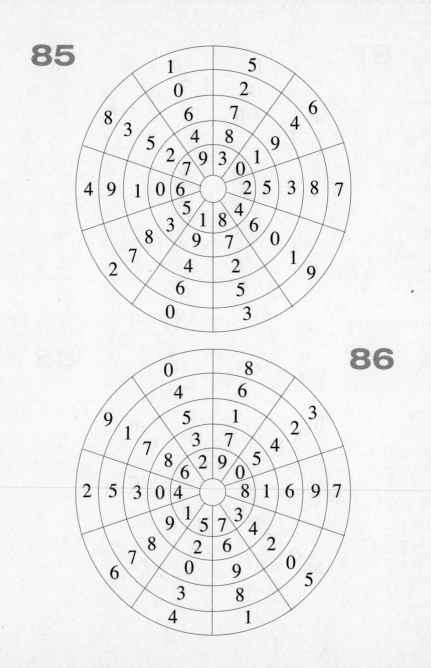

87

88

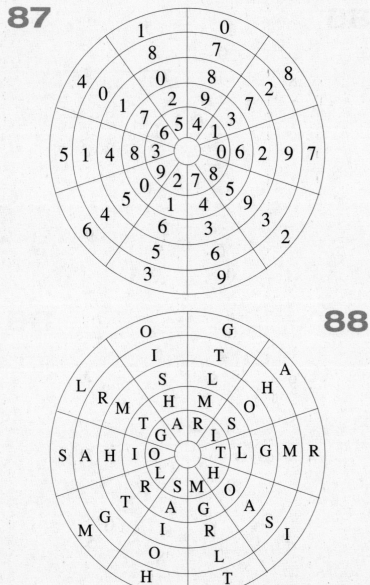

89

90

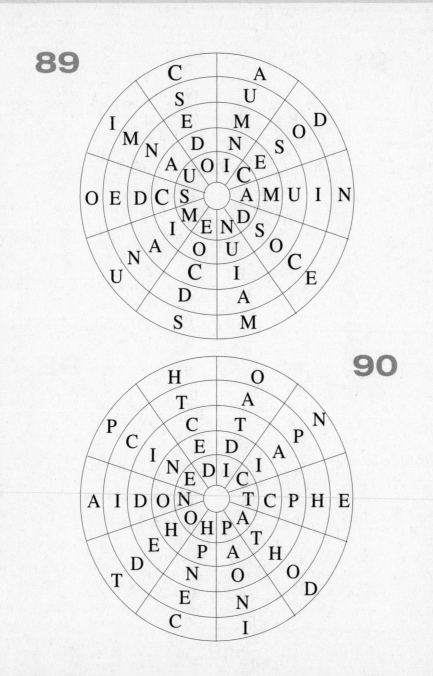

91

92

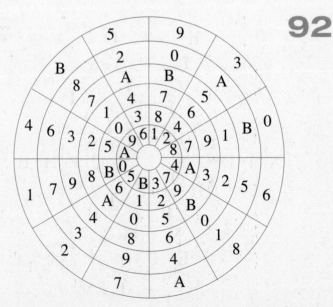

93

94

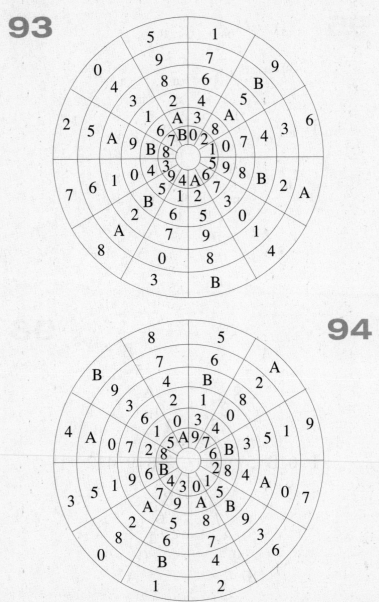

95

96

97

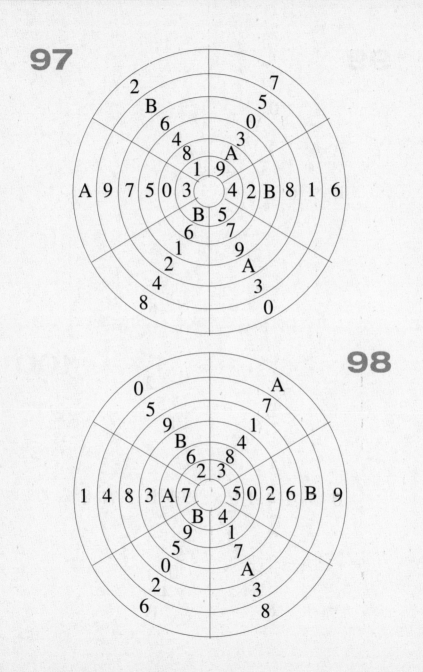

98

99

100

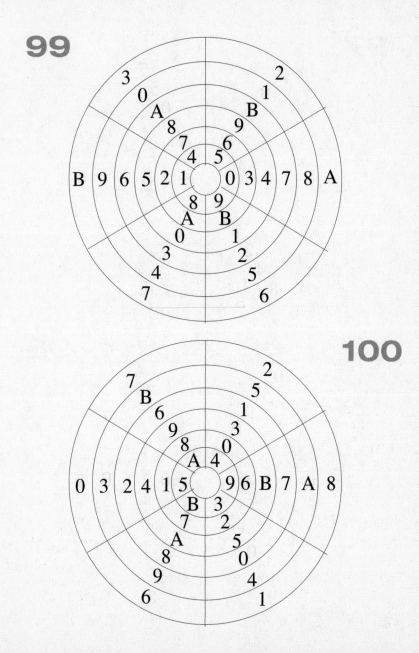

101

102

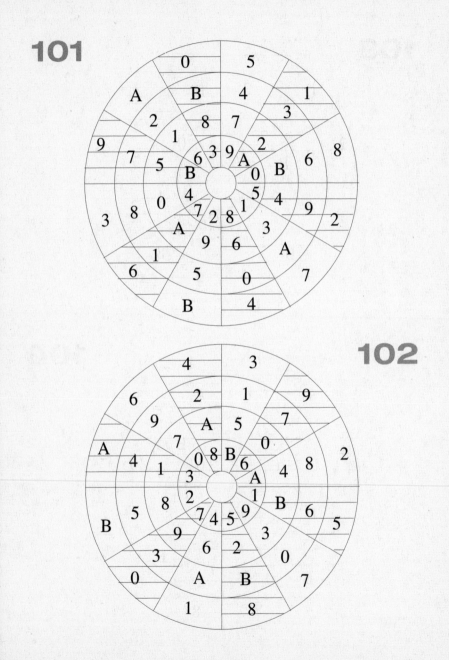

103

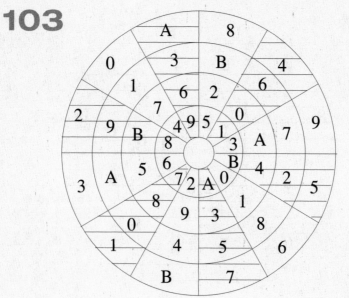

104

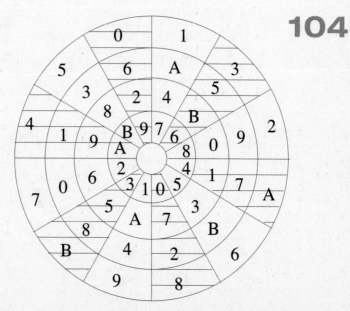

105

106

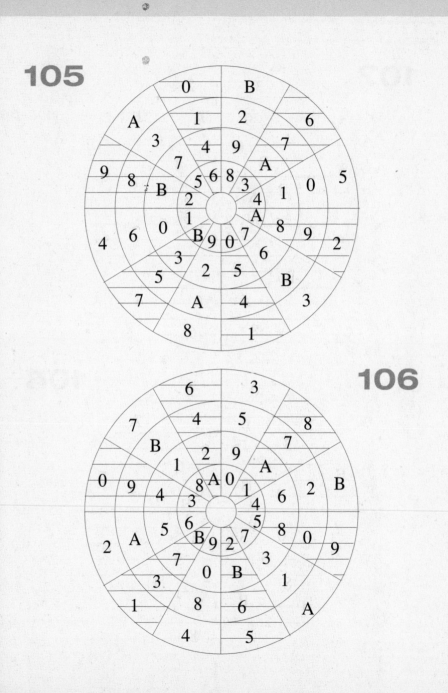

107

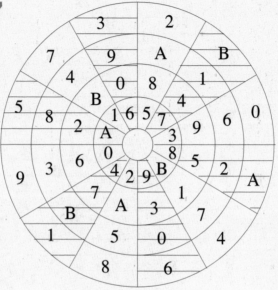

108

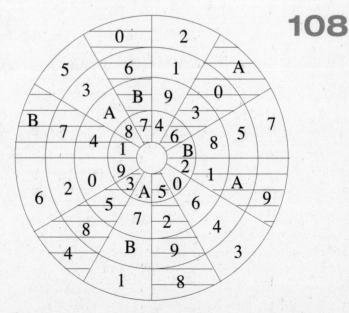

109

110

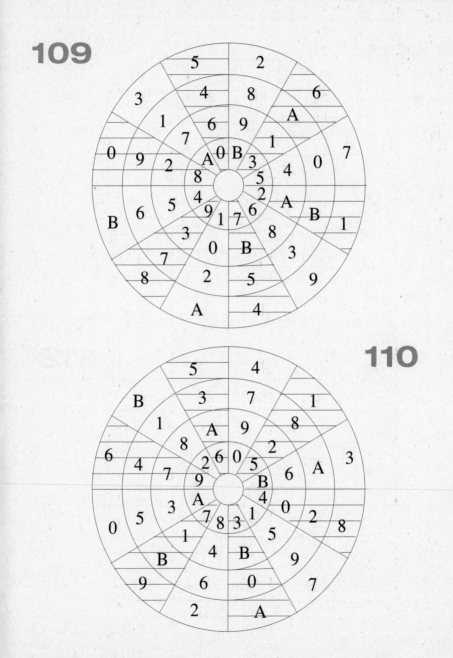

111

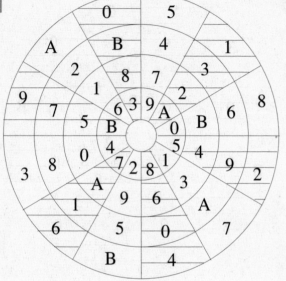

112

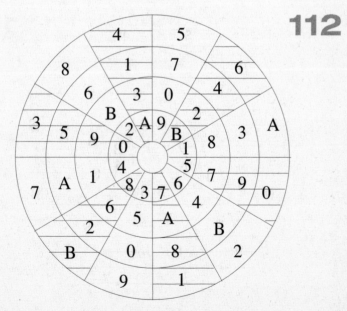

113

114

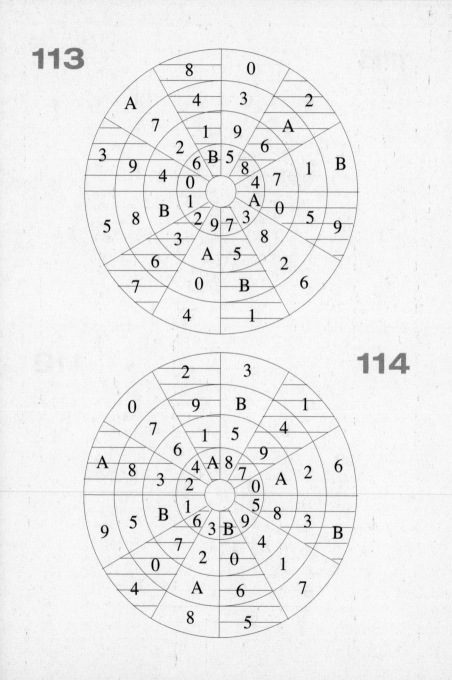

115

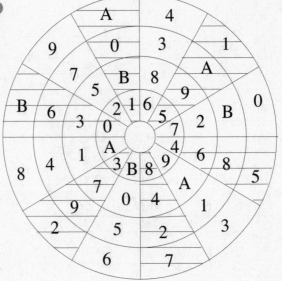

116

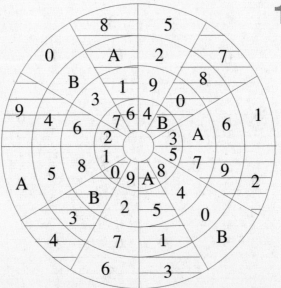

117

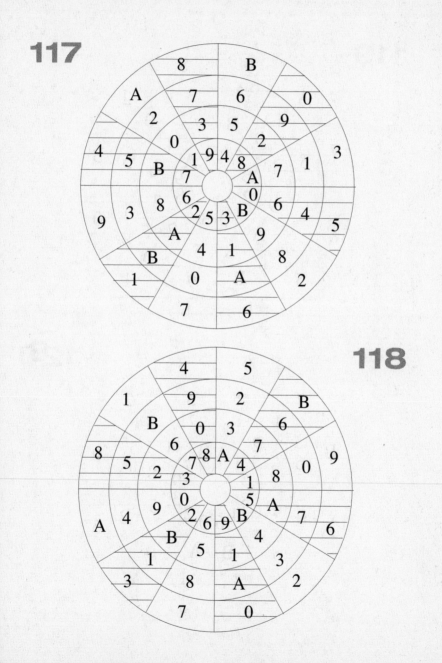

118

119

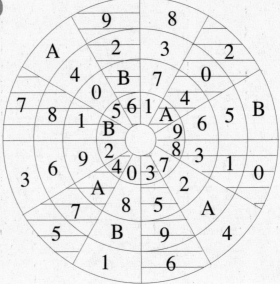

120

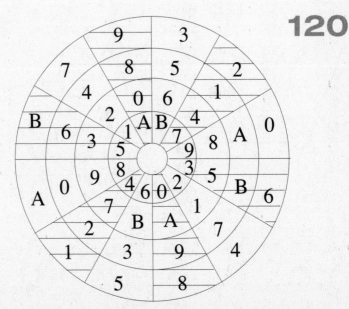

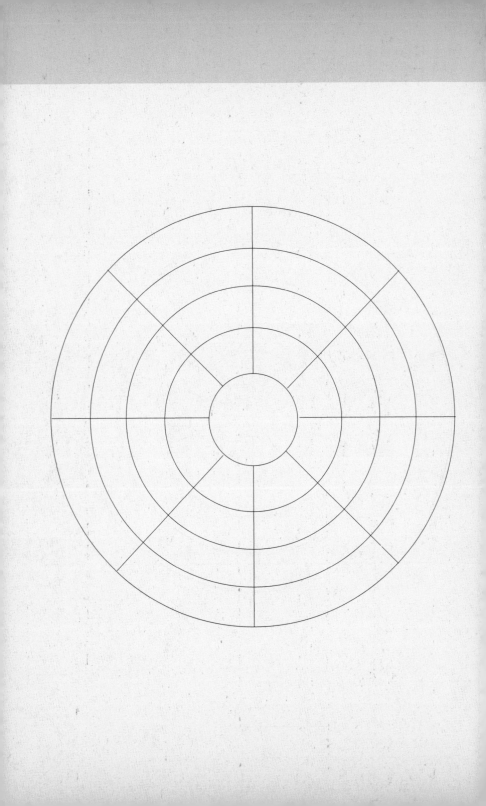

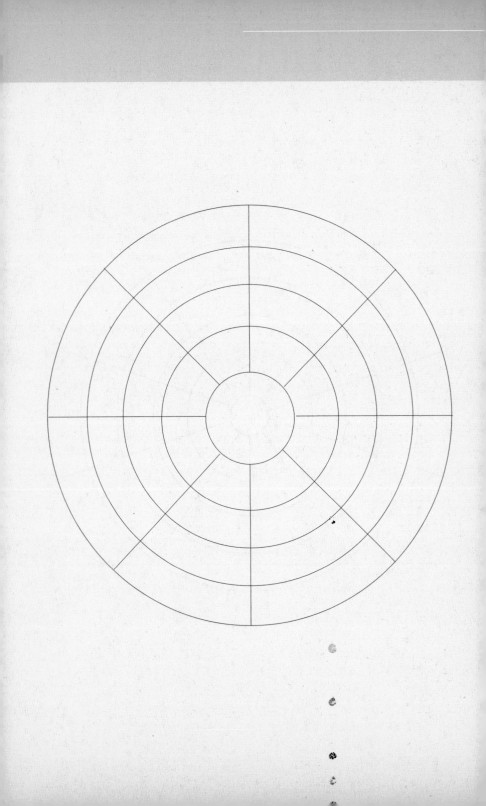

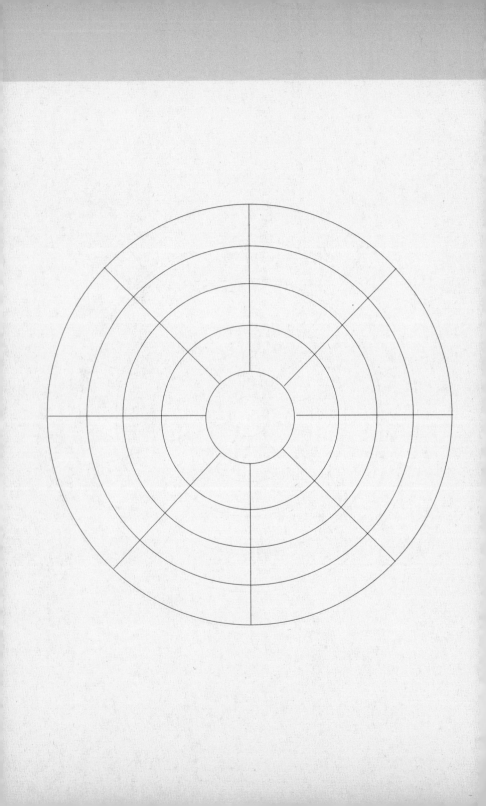